Biology in Physics

Is Life Matter?

Series in *Polymers, Interfaces, and Biomaterials*

Other books in the series:

Alexander Grosberg, *Theoretical and Mathematical Models in Polymer Research: Modern Methods in Polymer Research and Technology* (1998).

Kaoru Tsujii, *Chemistry and Physics of Surfactants: Principles, Phenomena, and Applications* (1998).

Teruo Okano, Editor, *Biorelated Polymers and Gels: Controlled Release and Applications in Biomedical Engineering* (1998).

Also Available:

Alexander Grosberg and Alexei R. Khokhlov, *Giant Molecules: Here, There, and Everywhere* (1997).

Jacob Israelachvili, *Intermoleculer and Surface Forces, Second Edition* (1992).

Biology in Physics

Is Life Matter?

Konstantin Bogdanov
Institute of Developmental Biology
Russian Academy of Sciences
Moscow, Russia

ACADEMIC PRESS
A Harcourt Science and Technology Company
San Diego San Francisco New York Boston
London Sydney Tokyo

This book is printed on acid-free paper. ∞

ACADEMIC PRESS
A Harcourt Science and Technology Company
525 B Street, Suite 1900, San Diego, CA 92101-4495, USA
http://www.apnet.com

Academic Press
24-28 Oval Road, London NW1 7DX, UK
http://www.hbuk.co.uk/ap/

Library of Congress Catalog Card Number: 99-65060
International Standard Book Number: 0-12-109840-0

Printed in the United States of America
99 00 01 02 03 EB 9 8 7 6 5 4 3 2 1

To my parents

Contents

Foreword

"The whole of science is nothing more than a refinement of everyday thinking," A. Einstein. However, along with the development of human society, both the weight and the role of different sciences have varied. In the seventeenth and eighteenth centuries mechanics was focused upon, including heavenly mechanics, but in the nineteenth century it is already difficult to establish whether it was physics (and mechanics) that played the leading part, or whether it was chemistry or biology. But from the end of the nineteenth century and until the middle of the twentieth century, physics domineered; it was the top priority for all scientists. In 1897 the electron was discovered; soon after; radioactivity and x-rays; in 1900 the quantum theory appeared, and the whole physical outlook of the world changed. The development of physics reached its climax point, first with the special and general relativity theories (1905 and 1915), and then with quantum mechanics (in the 1920s). It was then that the atom and atom nuclei structure were clarified, and as a result, modern natural sciences, including chemistry and biology, started their independent development. Unfortunately, nothing but physics was the crucial factor for both the A-bomb and the H-bomb creation. The era of physics was over in the 1950s — exactly at that period of time biology suddenly rushed forward, when DNA structure and the nature of genetic code were described (1953). Thus, the development of molecular biology was triggered.

Now that we are entering the twenty-first century, it is undoubtedly biology that plays the leading role, as far as the interest of the human society and the development of its potential are concerned. Illustrating this idea very well is one of the most famous international scientific journals, *Nature*: its basic weekly issue covers all sciences, biology among them. As for its six monthly satellites, all of them contain articles only on biology or areas related to biology: *Nature: Genetics*, *Nature: Structural Biology*, *Nature: Medicine*, *Nature: Biotechnology*, *Nature: Neuroscience*, *Nature: Cell Biology*.

Of course, the current progress in biology would be impossible without modern physical devices and methods. In this respect, particularly in technology and computer science, physics has not shifted to the background. On the other hand, biology provides new food to physics, particularly when it offers examples of information control and transfer. Examples of this kind show how biology is valid for physicists, and how physics is important for biologists. This book by Konstantin Bogdanov spans the two sciences, and should be equally interesting for both physicists and biologists.

In conclusion, I would like to touch upon a problem of primary importance, that of reduction: Is it possible to explain biology through physics? Since we know the laws of the electron and atom nuclei interaction that form the matter of living organisms, we should be able to explain all the living organisms' processes from the point of view of physics. A lot has been achieved by scientists working in this direction, and yet there is one thing that remains unclear: the origin of life, the step between life and nonlife. Perhaps the molecular approach will help to clarify the transformation from complex molecules to simplest organisms that can self-regenerate. But can physics explain the emergence of conscience? I personally don't understand it at all. Those who believe in God solve this problem very easily: nonliving matter becomes live when God "inhales" life into it; and it is also God who supplies a human being with a conscience. Unfortunately, this explanation does not convince atheists, me among them; it only substitutes one unknown for another, and is definitely beyond the scientific approach and scientific outlook. Nevertheless, if we remain within this approach, we cannot consider the reduction of biology to modern physics fully proven. What if there are other fields and particles that have not been yet discovered by physicists as well as processes valid for living organisms only, and not for nonliving ones? Of course, this assumption does not reject reductionism in principle, only the reductionism based on what modern physics knows at present. To be honest, I am saying it only to be careful. Most likely, at the fundamental level (fields and particles) no new discoveries in physics are necessary to explain what happens in biology. On the other hand, the physics of complex systems, to say nothing of living organisms, is still quite vague. To solve the problem of reductionism, that is, what connects basic biology with physics, is, to my mind, the central problem of twenty-first century science.

Of course, Konstantin Bogdanov's book is rather far from problems of this kind; yet it is indirectly connected with them and can facilitate the interaction of physicists and biologists, which will enable more rapid progress in biology.

The objective of the book is to make biology more attractive to physicists. Still it cannot be called popular, as the reader has to have a certain academic background in physics. Every chapter triggers further research and contains comprehensive reference to recent publications.

This is not a textbook; rather, it is a collection of separate stories about how physics can be used when biology is studied. The author does not try to categorize the results of biophysical research. This is why what you feel after you have read this book is similar to what you feel in an art gallery. It is quite likely that this method of research in biology can become more attractive, particularly for the physicists who find it difficult to approach complicated physiological processes.

This book is very useful for university students and physicists as well as for anyone who is still fascinated by the perplexing script according to which biological processes are carried out in both ourselves and nature around us.

Vitaliy L. Ginzburg
Member of Russian Academy of Sciences
Lebedev's Physical Institute
Moscow, Russia

November, 1998

Acknowledgments

Production of this book would not have been possible without the help of many people. However, the first and most honorable position in this list of acknowledgments must be given to my first teacher in biophysics, Dr. Vladimir Golovchinsky, who introduced me to biophysics three decades ago when I was still an undergraduate at the University of Physics and Technics at Moscow. Special thanks to Dr. Yuliy Brook who gave me the idea to write the book many years ago, and Dr. Alexander Grosberg, who helped me choose the most appropriate style of presentation. I am particularly grateful to Dr. Alexei Chernoutsan who found and recovered the manuscript of my book considered as lost forever. It has been a pleasure working with Mrs. Maria Ovchinnikova, whose skillful work as an artist created a handsome volume from an author's dream. I gratefully acknowledge the various authors and publishers who gave their permission to reproduce several figures and tables. Many thanks to Dr. Zvi Ruder for help and support in preparation of the manuscript for publishing. Finally, I am sincerely grateful to academician Vitaliy Ginzburg for writing the foreword for my book.

I promised my wife, Nadejda Kosheleva, that I would not give her the stereotypical public thanks for her support, and I am not doing so! Also, I feel sorry that in the very beginning I forgot to thank Dr. Edward G. Lakatta who taught me to run wishbone, helping me to stay healthy and wealthy while writing this book.

Introduction

Life is much too important
a thing ever to talk seriously about it.
Oscar Wilde, *Vera, or The Nihilists*, 1883

We are entering the third millenium. Of course this statement is relative: we merely have to deal with decimal notation. If we were using a hexadecimal system to count how many years have passed since Jesus Christ was born, we would have another 2096 years to wait for the third millenium. However, is it the only convention produced by the human mind? The answer is no.

"Living" and "nonliving" nature is another example. Can we really distinguish between living and lifeless objects? Since conventions generate one another, for many years different people studied living and nonliving nature separately. Physicists, biologists, and chemists were looking for laws common for nonliving objects, and managed to accomplish a great deal. Meanwhile, biologists were producing piles of data concerning "live" nature that can hardly be systematized.

Biological laws are so numerous and complicated that physicists used to feel scared: What would happen to them if the law governing the motion of a stone thrown at a certain angle to the horizon depended on the shape and size of the stone? Or, can you imagine different laws for the motion of bodies thrown up and down? Yet, unfortunately, that is what is going on in biology. For example, however thoroughly we may have studied the law of the rat's heart systole, we find it next to impossible (or impossible) to apply it to a human heart. Likewise, it is necessary to use different equations to describe the process of contraction and relaxation of a muscle, whereas for a steel spring one is enough. Probably that is why the number of publications in biology is a dozen times greater than that in physics as you can see by looking through *Science*, *Nature*, or *Proceedings of the National Academy of Sciences*.

However not all the physicists have been scared off by the complexities of biology. Some, who call themselves "biophysicists," decided to help biologists and crossed the border between the alive and the lifeless. As they stepped over to the other side, the first thing they saw was that biologists speak a language they don't understand. The most interesting thing was that this language was familiar to physicists—a kind of broken English—but there were plenty of absolutely unintelligible terms (see the index at the end of the book). Like all aliens, biophysicists have rather mixed feelings as they invade the unknown country of biology. Though scientific curiosity incites them, they realize that it is impossible to learn much without good command of the language.

For a long time I had been among such aliens, until I compiled a phrase-book intended to help me communicate with biologists in their language. It is this phrase-guidebook that is offered here. The book is divided into a number of chapters concerning different, rather unrelated, fields of biology. In each chapter you can find the description of some physical law applied to the explanation of specific biological phenomena, which is done to remind you of the native language of physics that you spoke before crossing the boundary.

Every tourist usually tries to decide where to go. A big city or wilderness? Of course, in a city you find good service and a lot of sophisticated entertainment, but the impressions to share with friends will hardly come as a surprise to listeners. Besides, it is unlikely that urban tourists could make any real discoveries. Now, unexplored paths is an entirely different matter—they make you a pioneer. But how can you get there? And is it really worth the effort? It may very well turn out that you find nothing interesting as you reach the place.

Anticipating such questions, Table 0.1 lists large biological "cities" frequently visited by physicists who work in Mechanical and Electrical Engineering departments. The table was created as a result of the analysis of scientific articles found in the database MEDLINE (U.S. National Library of Medicine), published in 1997. Only the articles where the first author worked in Mechanical or Electrical Engineering departments are included. This table provides the answer to another question: In what area of biology can a physicist apply his or her knowledge and still remain a research team leader?

Table 0.1 Biological issues considered in papers performed in Electrical and Mechanical Engineering departments and published in 1997.

Issues considered	Number of papers published	% of total
Departments of Electrical Engineering, 1997		
Tomography, Magnetic Resonance Imaging—Methods-analysis	36	20
Electric stimulation—methods	28	15
Ultrasonography—methods-analysis	19	10
Cardiography—methods-analysis	19	10
Neural Networks Computer-diagnosis	17	9
Neurology—methods-analysis	17	9
Physiology of heart and circulation—models	13	7
Others	35	20
Department of Mechanical Engineering, 1997		
Mechanics of Bones, Joints, Limbs and Human Body, Prothesis	79	48
Mechanics of Blood Circulation	40	24
Cryosurgery & Cryopreservation	12	7
Minimally Invasive Surgical Procedures, Robotics	8	5
Others	25	15

If you really are fond of travelling and don't like conventions, welcome to a mysterious country — Biology.

Biology in Physics:

Is Life Matter?

1

Electricity Inside Us

It is hard to imagine what would happen if we were suddenly deprived of electricity. A global catastrophe, which could destroy thousands, or even millions, of human lives, could very well inspire Stephen Spielberg to make another movie called "Lights off, please." Certainly the people who would suffer the least are those beyond civilization, like the Amish, who live in southeast Pennsylvania and who reject electricity for ideological reasons. Of course, it is possible to de-energize the houses, to live without television, radio, telephone, and PC (as the Amish do), but all the same we cannot be completely "saved" from electricity since electricity is inside us.

For more than four centuries scientists have been attempting to define the role of electromagnetic phenomena in the life of humans and animals. But only after Hans Christian Oersted discovered that the electric current flowing through a wire loop can deflect a magnetized needle, did it become possible to create sufficiently sensitive galvanometers. With the help of these galvanometers, Carlo Matteuchi (1838), and later Emil du Bois-Reymond (1848), measured the electrical fields generated by the contraction of the muscles of animals and humans. However, living organisms are not only generators of electricity, but have high sensitivity to an electromagnetic field, whereby it is in no way possible to explain the observed effects by the thermal action of such a field.

It is well known, for example, that the general anesthesia (the loss of consciousness and of the sensitivity to pain) can be obtained by passing the impulses of alternating current through a human brain, which is frequently used for anesthesia during surgical procedures. The direction of the lines of force of the electric field of the earth serves as a compass for long-distance migrations of the Atlantic eel. The navigation abilities of pigeons are based on the perception of the magnetic field of the earth. Skeletal bones inside an electric field grow in a different way, which theoretically can be used for treatment of fractures. The list of the biological effects of an electromagnetic field can be continued, but it is not our task.

Luigi Galvani and Alessandro Volta

Luigi Galvani, Professor of Anatomy from Bologna University, was the first to discover the effects of an electric field on a living organism. Since 1775 he was interested in the connection between electricity and life. In 1786 one of the professor's assistants, while separating a muscle from a frog's leg with a scalpel, happened to touch the nerve going to this muscle. At this very moment a static electricity generator was rotating on the same desk. Each time the electric machine produced a discharge, the frog's muscle contracted. Galvani concluded that the electricity would somehow *go* into a nerve, which would result in a muscle contraction. He devoted the following five years to the study of how different metals can cause muscle contractions. Galvani came to the conclusion that, if a nerve and a muscle lay on identical metallic plates, the shorting of plates by a wire did not give any effect. However, if the plates were made of different metals, their shorting was accompanied by a muscle contraction.

Galvani reported his discovery in 1791. He thought that the reason why a frog's leg jerked was the "animal electricity" generated inside an animal's body, whereas the wire only provided for the closing of an electric circuit. He sent a copy of his work to Alessandro Volta, the physics professor from Pavia (northern Italy).

Alessandro Volta repeated the experiments of Galvani, and obtained the same results. At first he agreed with Galvani's conclusions but then he paid attention to the fact that the animal electricity emerged only when there were two different metals in a circuit. Volta showed that when two different connected metals are placed on a human tongue, it feels like tasting something. Likewise, if you quickly touch your eyeball by a tin plate while having a silver spoon in your mouth, the shorting of the spoon and plate will give light sensation. Trying to refute Galvani's thesis about the existence of animal electricity, Volta suggested that a circuit that consisted of two different metals, both in contact with a salt solution, could be a source of direct current, unlike an electrostatic machine, which could only produce electric discharges.

This proved to be true. In 1793 Volta published his paper containing the description of the first source of direct current (later called *galvanic*). Although soon after, Galvani showed that animal electricity exists in the electric circuits without bimetallic contacts as well, he could not continue his dispute with Volta. In 1796 Bologna became a French territory, and Galvani, who refused to recognize a new government, lost his position in the university. He had to look for refuge at his brother's place, but he did not practice science again, and died in 1798. In 1800 Volta presented his discovery to Napoleon and received a great reward for it. Thus, the dispute of two compatriots who had different spirit, education, and political views, triggered the development of modern physics and biology.

Who was right in this dispute? Does animal electricity exist or not? In his latest experiments Galvani used two muscles at a time arranging them in such a way that the nerve leading from one muscle was placed on the other muscle (Figure 1.1).

It turned out that at each muscle-1 contraction caused by the current passing through its nerve, muscle 2 also contracted, as if current passed through its nerves as well. From these experiments Galvani concluded that a muscle at the moment of contraction served as a source of electric current. So it was demonstrated (though, in an indirect way) that the animal electricity exists. But it was not until half a century later, in 1843, that the German physiologist Emil du Bois-Reymond demonstrated the presence of the electric fields in nerves and muscles, using the latest available electro-measuring instruments. It is interesting that he was assisted by Werner Siemens and Georg Halske, who were hardly known at the time. Later, in 1847, they founded the telegraph company Siemens and Halske, which soon developed into a famous industrial empire.

So what is the source of the animal electricity? It took another half a century to answer this question.

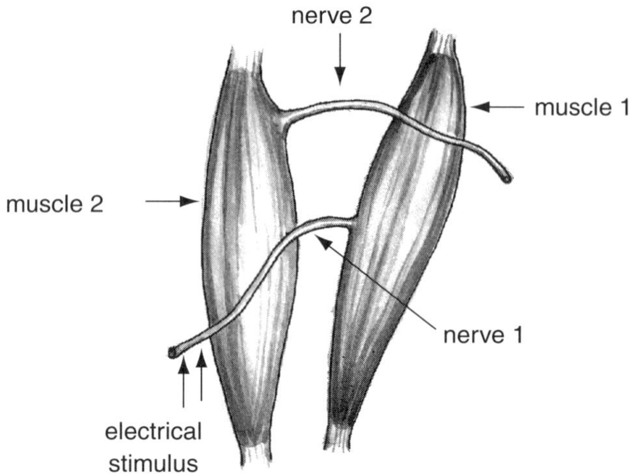

FIGURE 1.1. The set-up of L. Galvani's experiment to prove animal electricity.

Cell Membrane: Lipid Bilayer and Ionic Channels

We consist of cells as a house consists of bricks. The analogy cannot be further developed because we consist of hundreds or even thousands of types of "bricks". The cells of skeletal muscles differ from nerve cells, and red blood cells seem to have nothing to do with cells of blood vessels. Nevertheless, there is a property which makes all the cells alike: all of them are enclosed by a membrane.

The main feature of a living organism is to be "picky" in its relations with the environment. What helps here is the selective permeability of the membranes of the cells. The cell membrane is its skin, which is about 5 nm wide. The cell membrane selectively reduces the speed of molecules moving into a cell and out of it. It defines which molecules are to penetrate a cell, and which are to remain beyond its limits. Thus, the function of a cell membrane is in many respects similar to that of security for a foreign embassy building. A high fence around the building and several entrances where security people decide who can be let inside and who can't usually serve this purpose.

The role of a "high fence" in a cell membrane is played by the phospholipid bilayer (Figure 1.2), which forms its basis and makes it impervious for most water-soluble molecules. Phospholipid is a very long molecule, which has a polar head group and two hydrophobic hydrocarbon tails whose length varies from 14 to 24 carbon atoms. Phospholipid molecules stick to each other in water by hydrophobic tails forming a bilayer, so that their hydrophylic polar heads, shown as beads in Figure 1.2, remain in contact with water molecules.

Different ions moving along the gradient of concentrations can cross a cell membrane only where it has *pores* or *channels* specifically intended for crossing over. The ionic channels (shaded in Figure 1.2) are big protein molecules that create, in the lipid bilayer, pores for water and for the vitally important ions sodium, potassium, calcium, and chlorine. Some parts of these protein molecules are charged, therefore they can move due to the changes in the electric field, thus altering the configuration of the entire ionic channel. As a result, the channel conductivity changes. In Figure 1.2 these charged parts of the ionic channel are schematically shown as gates that open (activation) and close (inactivation) the gap of the ionic channel.

There are a lot of ionic channels that differ from one another by their features. As a rule, each ionic channel is adapted to let only one of the previously listed ions through the membrane. Such channels are called selective potassium, sodium, etc., depending on which ions they allow to pass. However, there are also channels that let several types of ions through, called nonselective. The density of ionic channels is uneven, ranging from several units to several thousand units per $10^{-6}\,mm^2$ of a membrane.

Though the molecular structures of many channels' proteins are already well-known, the mechanisms of selective permeability of these channels with respect to different ions are still not clear. The most reasonable explanation of the ionic channel selectivity seems to be the different size of ions. But then how can a selective channel for a large ion be created? Evidently, smaller ions will also pass through a selective channel for a larger ion. Anyway, what does a size of an ion mean?

Water is a strong polar solvent. Therefore, each ion in the aqueous solution is enclosed within a shell of several water molecules (hydration shell). How does the channel's protein recognize the type of ion if the latter is surrounded by a water shell? The answer is simple: Before entering the channel the ion partially undresses, and proceeds through the channel half-dressed. Therefore, the cross section of such a half-dressed ion may have very weird shapes. The protein channel, in its narrowest place (called the *selective filter*), is considered to have the cross section transmitting its own

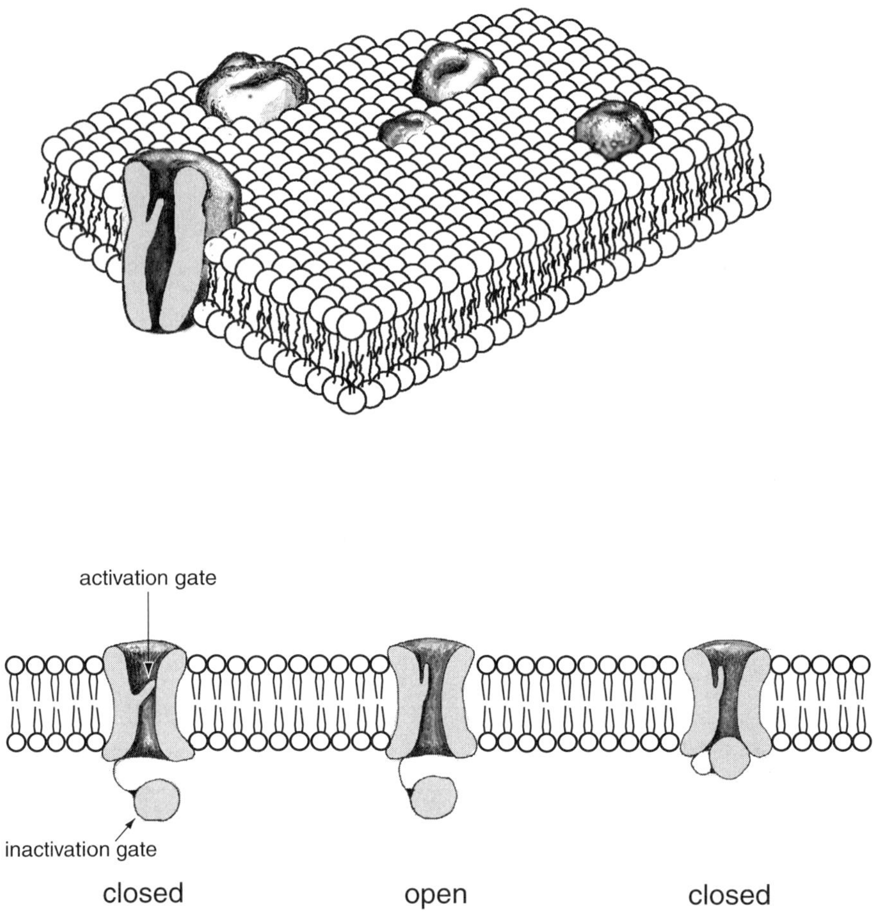

FIGURE 1.2. Biological membrane. *Top*: The general view of a phospholipid bilayer with ionic channels floating in it. *Bottom*: Schematic section through the ionic channels in different states, closed (activation gate is closed), open, and closed (inactivation gate is closed).

ions only. Thus, for example, according to the hypotheses of Dwyer *et al.* (1980), the selective filter of the potassium channel must have a circular cross section of 0.33 nm in diameter, and the selective sodium filter, a rectangular cross section of 0.31×0.51 nm.

Some ions can move inside a cell even when their concentration outside it is less than inside. It happens because of the presence in the lipid bilayer not only of the channel proteins, but also of the proteins that act as ion pumps.

Table 1.1 Concentration of some ions inside and outside of cardiac myocytes.*

Ion type	Concentration, mM/l	
	inside	outside
Na^+	10	140
K^+	150	5
Cl	20	145
Ca^{2+}	< 0.001	2
Organic anions	~120	—

*The data are taken from comparing the ionic composition of intracellular and extracellular solutions in the experiments with the voltage clamp.

The activity of the membranous protein-pumps is accompanied by great energy consumption, and as a result ion concentration inside a cell can be dozens, or even thousands, of times higher (or lower) than that outside (see Table 1.1). For example, potassium ion concentration inside a cell is 30 times higher than in extracellular liquid. On the contrary, the sodium ion concentration inside a cell is 14 times lower than outside. As we can see, the differences in potassium and sodium ion concentrations on each side of a membrane are necessary for electric fields to exist in living organisms.

It turned out that at rest, a cell membrane is permeable only for the potassium ions. In other words, when muscle cells are not contracting and the nerves are resting rather than thinking, only potassium channels are opened in their membranes. On the other hand, when the same cells are active, or, to put it in other words, *excited*, their sodium and calcium channels are opened as well, increasing for a short time the concentration of respective ions inside a cell. So what is it that controls the channel activity?

Resting Potential

Let's try to imagine what might be the result of the difference in the potassium ion concentration on each side of a cell membrane with high permeability for these ions. (It was this problem that was posed and solved in 1902 by German physiologist Julius Bernstein, the author of membrane excitation theory.) Suppose that we immerse a cell with a membrane that is permeable only for the potassium ions into electrolyte, where their concentration is lower than inside the cell. Immediately after the membrane comes into contact with the solution the potassium ions begin to get out of cells, like gas gets out of an inflated balloon. However, each ion carries a positive electric charge, and the more potassium ions leave a cell, the more electrically negative its contents becomes. Therefore, each potassium ion coming out of cells will be affected by an electric force preventing it from leaving the cell. Eventually, the balance will be established, whereby the electric force acting on a potassium ion in the membrane channel will be equal to the force resulting from the difference of potassium ion concentrations inside and outside the cell. It is obvious that such balance between internal and external solutions will set up a potential difference. If the potential of the external solution is taken to be a zero potential, the potential inside a cell will be negative.

This potential difference, the simplest of the observed bioelectrical phenomena, is called the *rest potential* of a cell. As it was shown by W. Nernst in 1888, in an ideal case when a cell membrane is permeable only for the ions of a single type (S), a potential difference E_S is established on a membrane. This potential difference is called equilibrium potential for the given ion:

$$E_S = E_1 - E_2 = (RT/zeA)\ln([S]_2/[S]_1) \qquad (1.1)$$

where e is the electron charge, z is the valency of ion S, A is the Avogadro constant, R is the gas constant, T is the temperature in K, $[S]_1$ and $[S]_2$ are the concentrations of the ion S on each side of a membrane. Substituting $[K^+]_2/[K^+]_1 = 1/30$ and $T = 300$ in Eq. (1.1), we obtain $E_K = -86\,\text{mV}$, which is close to experimentally obtained values of a resting potential of many living cells.

It is worth mentioning that voltage drop on a cell membrane, which is less than 0.1 V, happens on a segment about 10^{-6} cm long. Therefore, the strength of the electric field inside a membrane can reach huge values— about 10^5 V/cm, which is close to the strength of the electrical breakdown of this membrane.

Thus, as a result of the ion channels existence and the great difference in concentration between intracellullar and extracellular solutions for many ions (see Table 1.1), potential difference on cell membranes appear. If channels for one ion only are opened in a membrane, the potential difference on it will be equal to the equilibrium potential for this ion. But what will happen if the membrane is permeable for several types of ions at once?

Goldman (1943) and Hodgkin & Katz (1949) deduced the formula for a potential difference, E, on a membrane permeable for several ions at once assuming that the electrical field inside a membrane is uniform:

$$E = \frac{RT}{zeA} \ln \frac{P_{\mathrm{Na}}[\mathrm{Na}]_e + P_{\mathrm{K}}[\mathrm{K}]_e}{P_{\mathrm{Na}}[\mathrm{Na}]_i + P_{\mathrm{K}}[\mathrm{K}]_i} \tag{1.2}$$

where [K] and [Na] are the concentration of potassium and sodium ions, respectively, inside a cell (with the index i) and outside a cell (with the index e); P_{K} and P_{Na} are the permeabilities of a membrane for these ions; and the remaining symbols are the same as in Eq. (1.1).

It is worth mentioning that the membrane permeability for a specific ion is expressed as the ratio of a diffusion coefficient of this ion in a membrane to its width. It follows from Eq. (1.2), known as the *constant field formula*, that when there are several types of ionic channels a membrane potential depends on the relative permeability of these channels. In other words, the higher the membrane permeability to any ion, the closer the membrane potential to the equilibrium potential for this ion; the potential is calculated using Eq. (1.1).

It is not that simple to measure the difference of electric potentials for living cells: the cells are very small. Because regular probes attached to every voltmeter can't be used here, glass pipettes (microelectrodes) are used, with a tip diameter less than one micrometer. The pipette is filled with a strong electrolyte solution (for example, three-molar solution of potassium chloride). The contents of the pipette are then connected through a metallic conductor with the input of a voltmeter with a high (more than 10^9 Ohm) resistance since the pipette resistance sometimes approaches 10^8 Ohm. The skill of a microsurgeon as well as special micromanipulators are necessary here to insert a microelectrode into a cell, still keeping it alive (Figure 1.3). The problem is that as soon as the opening in a membrane appears, the ion concentration inside a cell will be approaching its values for the extracellular medium; that is, sodium and calcium ions will be entering the cell, and potassium ions will be leaving it. This will cause the inevitable death of the cell. But even if we were lucky enough to insert a microelectrode inside a cell without killing it, while measuring the potential difference on each side of a

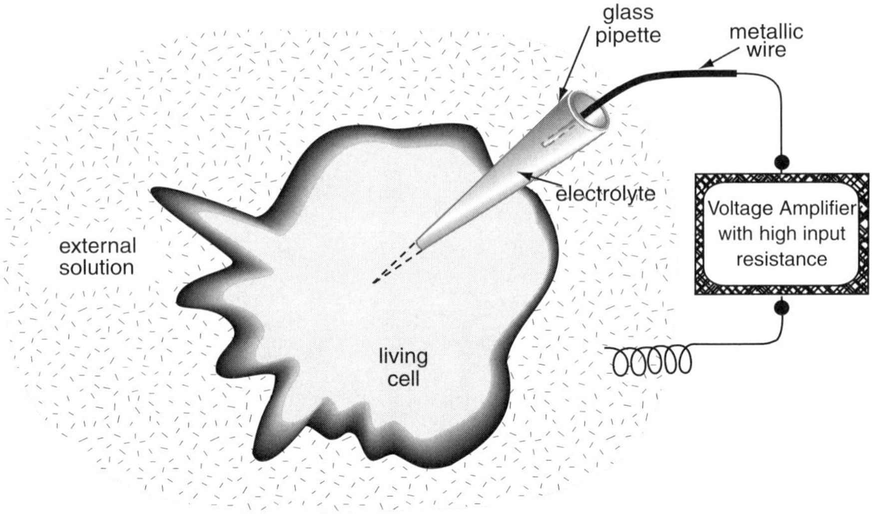

FIGURE 1.3. Measurement of the potential difference across the membrane of a living cell.

membrane we won't be able to keep track of the ions motion in a separate channel. Instead, we will see the result of the activity of millions of different ionic channels.

To follow the ions movement through single channels not only is the skill of a microsurgeon necessary, but also electronic instruments with extremely low levels of interior noise, as the value of electric current through a single channel often does not exceed 1 pA. Figure 1.4 shows the technique (patch-clamp recording)[1] that makes it possible to register the conductivity of a single channel and to record the changes of this conductivity in time. The first thing that attracts the attention is that at a constant voltage applied to a channel, the current passing through it changes in steps. Starting at a zero, when the channel is closed, it reaches its maximum value in the open state. Thus, the channel activity is of random nature, and we can only estimate the probability of its opened and closed states.

[1] In 1991 the German scholars Erwin Neher and Bert Sakmann were awarded the Nobel Prize for the development of this technique and for the analysis of the functioning of single ionic channels.

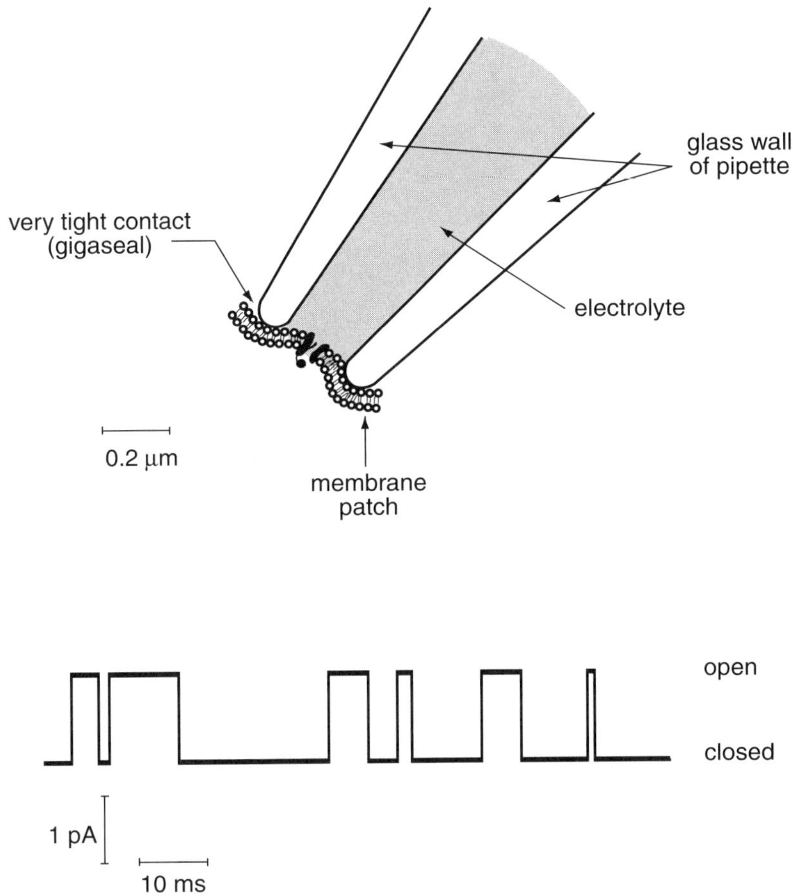

FIGURE 1.4. The patch-clamp technique used to record the conductances of a single ionic channel. *Top*: cross-sectional view of a glass pipette filled with electrolyte; its opening is closed with a piece of cellular membrane that contains just one channel. *Bottom*: record of the current through an ionic channel; *closed* corresponds to zero current through the channel.

Action Potential

What precedes the contraction of a muscle cell? What happens in nerve cells when we react to a prick of a needle? What is the excitation of cells? What is surprising is that in all these cases the same process takes place—a short-term change of the potential difference on a membrane. As we already know, in the state of rest the intracellular medium is charged negatively with respect to the extracellular one. During the excitation, the electronegativity of the intracellular medium decreases for a very short time (from 1 msec up to several decimals of a second), and sometimes a cell even becomes positively charged with respect to the extracellular solution. This transient change of the potential difference is called *action potential*, and if it happens in nerve cells, it is called *nerve impulse* (Figure 1.5).

In 1963 A. L. Hodgkin and A. F. Huxley were awarded the Nobel Prize for the discovery of the nature of the action potential. What followed from their experiments performed on a giant axon of Loligo squid was as follows.

1. The initiation of the action potential results from a short-term increase of the membrane permeability for sodium ions.
2. The ionic permeability of the membrane is a function of two variables: membrane potential and time; therefore, if the potential on a membrane is fixed (voltage-clamp technique), the task can be reduced to the study of permeability dependence on time only.

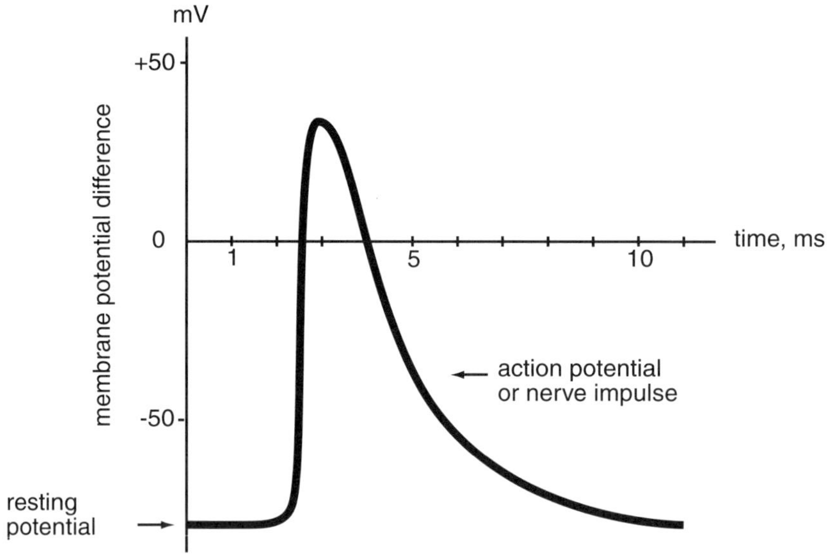

FIGURE 1.5. Change in the voltage across a membrane during a nerve impulse.

It turned out that the sodium channel of a nervous cell membrane differs from the potassium one. Its average conductivity (the probability of the open state) grows very quickly with the increase of the potential U of the intracellular medium which is counted from the potential of the external solution, taken as zero (Figure 1.6). Now let's suppose we are able to increase U by 20–30 mV (e.g., by passing the current through a cell). As soon as it happens, the sodium channel conductivity increases and, according to Eq. (1.2) U will increase even more, hence a sodium channel conductivity will increase too, and so on. Evidently, a small initial increase of U should trigger a fast explosion-like process, as a result of which the membrane permeability for the sodium ions increases to its maximum possible values and becomes dozens of times higher than its permeability for the potassium ions. The reason why it happens is that there are about 10 times as many sodium channels in a membrane as potassium ones. Therefore, if we disregard the membrane permeability for potassium, it is possible to evaluate potential U at the end of this fast process by using Eq. (1.1) and assuming $[Na^+]_2/[Na^+]_1 = 14$. After the substitution we obtain the potential leap during this transient process (equal to about 0.15 V).

However, the sodium channel has another feature peculiar to it, but not for the potassium one. Its conductivity depends not only on the voltage on

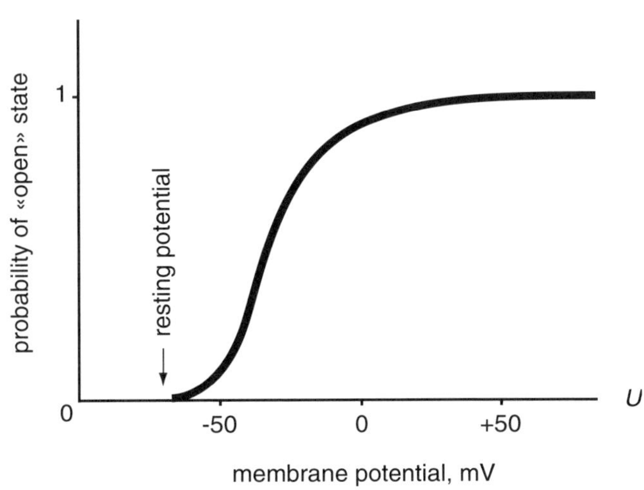

FIGURE 1.6. Dependence of the probability of the open state of the sodium channel on the voltage across a membrane.

the membrane, but also on how much time has passed since its opening. The sodium channel can be open only for 0.1 – 10 ms depending on the temperature and the kind of cell (Figure 1.7), which results in the return, after the sharp increase 0.15 V, of the potential difference on a membrane to its initial value, the resting potential. *U* returns to the resting potential even faster because the higher the conductivity of the potassium channels, the lower the sodium channel conductivity. The obtained results allowed A. L. Hodgkin and A. F. Huxley to propose a mathematical model that simulates the action potential.

Nerve Impulse Propagation

How do our senses inform the brain of what is going on around us? How do the different parts of our organism exchange information? For this purpose, nature devised two special communication systems. The first system, humor (from the Latin *humor*, meaning moisture, liquid), is based on diffusion or the transposition of biologically active substances with a liquid flow from where they are synthesized to the entire organism. This system is the only one that functions in protozoa as well as in plants.

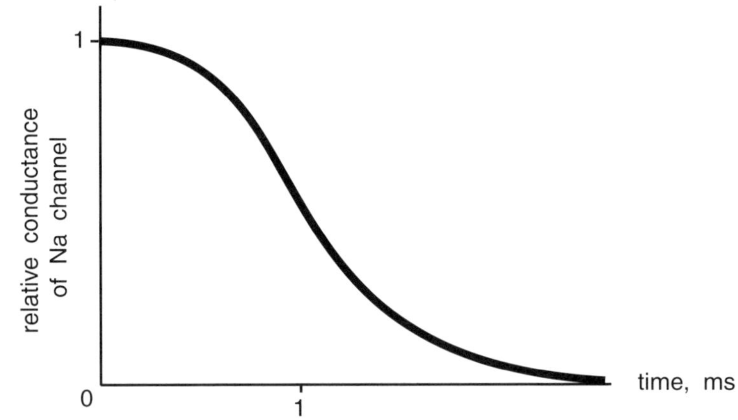

FIGURE 1.7. Inactivation of the sodium channel that occurs due to the closing of the inactivation gate (see Figure 1.2). Axis of ordinates: relative decrease of the sodium channel conductance upon attaining its maximum value; axis of abscissas: time.

Metazoans (including you and me) also have a nervous (from the Latin *nerves*, meaning sinew) system composed of a great number of nerve cells with appendices, the nerve fibers, running through the whole organism (Figure 1.8). The membrane of a nerve cell body is excited as soon as nerve impulses come to it from the adjacent cells through their appendices. This excitation passes over to a nerve fiber of this cell, and moves through it to adjacent cells, muscles, or organs at a speed of up to a hundred meters per second. Thus, an elementary signal transmitting the information from one part of an animal's body to another is a nerve impulse. Unlike the dots and dashes in Morse code, the duration of a nervous impulse is constant (about 1 ms), and the transmitted information can be encoded in a most peculiar way in the sequence of these impulses.

Quite a few of the well-known scientists of the past have attempted to explain the mechanism of excitation propagation along a nerve. In his famous book *Optics*, published in 1704, Isaac Newton suggested that the nerve had the properties of optical light guide. Therefore, "the ether vibrations" generated in a brain at will could propagate from there along solid, transparent, and homogeneous nerve capillaries to muscles, making them contract or relax. The founder of Russian science, the first Russian academician M. V. Lomonosov thought that the propagation of excitation along a nerve happened due to the movement of a special "rather fine nerve liquid" inside it. It is interesting that the velocity of propagation of excitation along a nerve was measured for the first time by a famous German physicist, mathematician, and physiologist Hermann Helmholtz in 1850, a year after A. Fizeau had measured the speed of light.

What makes it possible for the nerve impulse to propagate? What features of a nerve fiber define the velocity of propagation of an impulse through it?

To answer these questions, let us consider the electric properties of a nerve fiber. It is a cylinder whose lateral surface is formed by a membrane separating the inner electrolyte solution from the outer. This gives the property of a coaxial cable to a fiber, with the cellular membrane acting as insulation. But the nerve fiber is a very poor cable. The insulation resistance of this living cable is approximately 10^5 times less than that of a usual cable, as the thickness of the former is 10^{-6} cm, and the thickness of the latter is about 10^{-1} cm. Besides, the inner conductor of the living cable is an electrolyte solution whose resistivity is a million times higher than that of

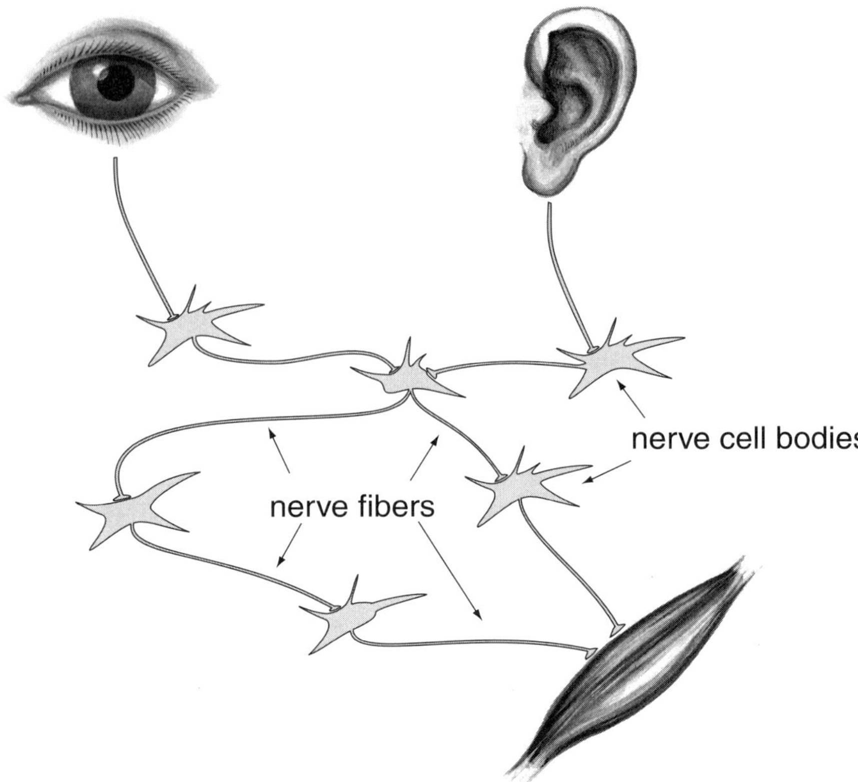

FIGURE 1.8. A considerably simplified diagram of connections between nerve cells, organs of senses, and muscles.

metal. That is why unexcited nerve fiber does not transmit electric signals over great distances too well.

It is possible to show that the voltage on a membrane of such fiber will exponentially decrease with a distance from the voltage source (Figure 1.9). The value λ used in an exponent, and which determines the decay rate of an electric signal in a nerve fiber, is known as the *space constant* of a fiber. The value of a space constant depends on a fiber diameter d, a unit area resistance of its membrane, r_m and the resistivity r_i of a liquid inside a fiber. This dependence has the following form:

$$\lambda^2 = (dr_m)/4r_i \tag{1.3}$$

Eq. (1.3) does not include the resistivity of the medium around the fiber as the dimensions of the external conductive liquid are mostly a lot larger than the fiber diameter, and it is possible to consider the outer solution equipotential.

Using Eq. (1.3) it is possible to find value λ for well-studied nerve fibers of a crab or a squid with $d \approx 0.1$ mm, $r_m \approx 1000$ Ohm·cm^2, and $r_i \approx 100$ Ohm·cm. The substitution of these values gives $\lambda \approx 0.2$ cm. It means that at a distance of 0.2 cm from the cell body, the amplitude of a nerve impulse must be almost three times less although the length of nerve fibers of these animals can be several centimeters.

However it does not happen, and the nerve impulse propagates along the whole fiber without reducing the amplitude. This is why it happens. As we saw earlier, 20–30 mV increase of the intracellular liquid potential with respect to the outer one causes its further growth and the emergence of a nerve impulse in the given region of the cell. From our calculations it follows that if a nerve impulse with the amplitude 0.1 V appears at the beginning of a fiber, then at the distance λ the voltage on a membrane would still exceed the resting potential by 30 mV, and a nerve impulse appears here as well; the same process takes place in the next segment of a fiber, and so on. Therefore, the propagation of an impulse along a nerve fiber can be compared with

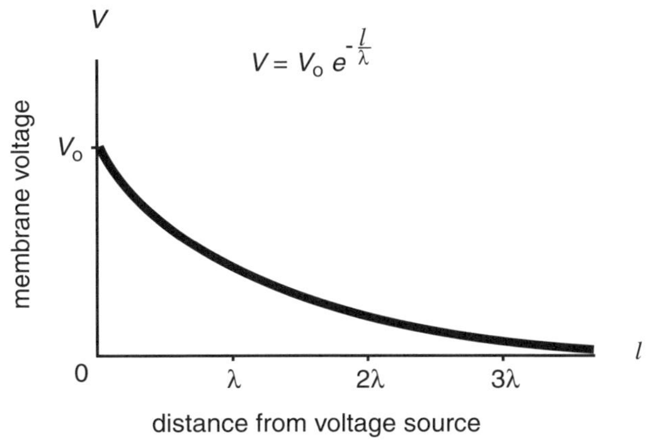

FIGURE 1.9. Dependence of the voltage across a nerve fiber membrane on the distance from the voltage source with its positive electrode inside the fiber and the negative one outside of it in the proximity of point $l = 0$.

flame spreading along Bickford's fuse; but it should be noted that with the former the required energy is supplied by the difference of concentration of potassium and sodium ions on each side of the membrane, whereas with the latter, it is supplied by the quick burning of easily inflammable insulation of the fuse.

It is obvious that the greater the value of the space constant λ, the faster the nerve impulse can propagate. Since the values of r_m and r_i are almost the same in different cells and animals, it turns out that λ, and consequently the impulse propagation velocity, must depend mostly on the fiber diameter and grow proportionally with the square root of its value. This conclusion completely coincides with the results of experiments.

A giant (about 0.5 mm in diameter) nerve fiber of a squid can show how nature has used the dependence of nerve impulse propagation velocity of the fiber diameter. It is well known that the squid running away from danger uses its "jet engine", forcing a great mass of water out of its mantle cavity. The muscular system contraction setting this mechanism in motion is started up by nerve impulses that propagate along several giant fibers. As a result, the high rate of response and the simultaneity of all this muscular system operation are achieved.

Nevertheless, it is impossible to use such giant fibers in every region of a nervous system where a high rate of reaction and the analysis of the incoming information are required, as they would take up too much space. Therefore, for more sophisticated animal organisms, nature chose quite a different way of increasing the propagation velocity of excitation.

Nodes of Ranvier

Figure 1.10 schematically shows a nerve fiber (cross section along the axis), most typical for our nervous system. This fiber is separated into segments, each being about 1 mm long. All such segments are covered by *myelin*—a fat-like staff with good insulation properties. Between the segments, at a place about 10^{-3} mm long, the membrane of the fiber is in immediate contact with the outer solution. The area where the myelin sheath disappears is called the node of Ranvier.

What should the structure of a nervous fiber entail? As it follows from Eq. (1.3) for length constant λ, when the unit area resistance (r_m) of the membrane grows, the value of λ, along with the impulse propagation velocity, should also grow. This enables the latter to become almost 25 times greater, compared to a nonmyelinized fiber of the same diameter. Besides, the energy loss for the propagation of excitation along a myelinized fiber is much smaller than that along a normal one since the total number of ions crossing the membrane is negligible in the former case. Thus, a myelinized fiber is a high-speed and economical communication channel in nervous system.

The study of ionic channels of a membrane is a new and rapidly developing field of biophysics. Although nobody counted the number of the discovered ionic channels, apparently it must be approaching the first hundred. For many channels even the genes responsible for the synthesis of corresponding proteins have been identified. The opportunity to apply new physical and mathematical approaches has attracted many physicists to this estate of biologists. If you choose to turn to biology at this crossroad, the best guidebook to lead you through the jungles of ionic channels might be Bertil Hille's book *Ionic Channels of Excitable Membranes* (1992).

A Menu for a Person Condemned to Death

If you have thoroughly read everything so far, you undoubtedly deserve some rest. There are a lot of fairly interesting "semi-scientific" facts, and to

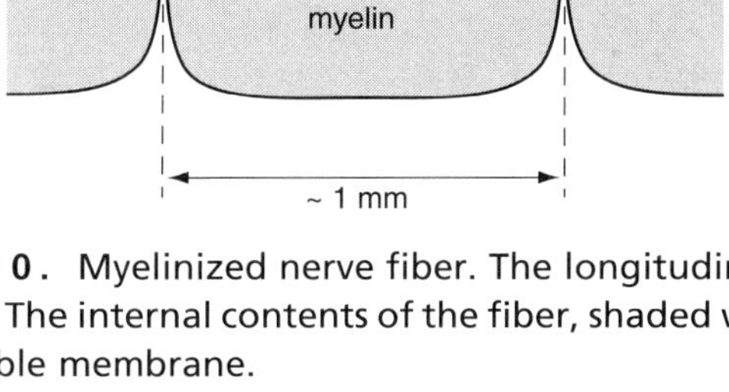

FIGURE 1.10. Myelinized nerve fiber. The longitudinal section of the fiber is shown. The internal contents of the fiber, shaded with dots, is enclosed with an excitable membrane.

study them helps you relax a great deal. As a rule, many of them begin with "It is well known that...".

Well, it is well known that a certain poison, tetrodotoxin, $C_{11}H_{17}O_8N_3$, completely blocks sodium channels. This substance seems to be the most low-molecular of all the known toxins of protein nature. This poison, which plugs a sodium channel like a cork plugs a bottle, was extracted for the first time from *Fugu* fish, or puffer fish, which inhabit the Sea of Japan. The poison was named after the Latin term for the fish of the Tetraodontidae family to which the puffer fish belongs.

Up to now, eating fugu's entrails has been considered in Japan to be a particularly refined way of suicide. Between 1927 and 1949, about 2700 men died of fugu poisoning. Selling this fish is forbidden in some regions of Japan; where it is allowed, only certified cooks are allowed to cook fugu.

Surprising as it may seem, the dish cooked with this fish is an exquisite delicacy of Japanese cuisine, despite the fact that some unfortunate gourmets died soon after. Though fugu has been notorious for thousands of years, there are still plenty of those who want to taste this dish.

Apparently, one of the first Europeans who had the privilege of tasting fugu was the English traveler James Cook, who was informed by the aborigines that this dish was poisonous—but not until he had eaten it. People say that when you are eating fugu, you don't only taste it—evidently the presence of what is left of tetrodotoxin in the dish gives a pleasant tingling and a warm feeling in the limbs, as well as euphoria. However, if you carelessly consume this dish, you can easily die because of the cessation of breathing.

Puffer fish uses its poison to scare predators away. In its skin, glands secrete tetrodotoxin when the fish is irritated by something. Tetrodotoxin can be found in the liver of this fish as well, and the scientists have yet to answer the question of why the nervous system of the fish remains immune to such potent poison.

The presence of tetrodotoxin was detected not only in fugu. It is also synthesized in certain species of salamanders, frogs, gastropods, crabs, and starfish. Some species of octopuses have tetrodotoxin secreted by glands located in suckers, so a "handshake" from such a creature is a real danger even for a human being. The only thing in common for different poisonous animals such as fugu, salamanders, frogs, and octopuses is that they all have tetrodotoxin contained in their spawn. Thus, it is obvious that the main function of the poison is to protect the offspring.

People learned how to use tetrodotoxin long ago. This deadly poison (it is 1000 times stronger than potassium cyanide) is mentioned even in fiction. For example, Agent 007 James Bond, a well-known character of British author Ian L. Fleming, was nearly poisoned by it.

Unfortunately, terrorists are using it now, and it is by no means a tale. This poison has an interesting property that is appealing to unscrupulous people. If the dose of tetrodotoxin is slightly less than a lethal one, a person gets into a state that absolutely looks like death. However, unlike real death, this state is reversible, and in a few hours a "dead" person comes back to life. Those who would like to know more about tetrodotoxin, fugu, poisoned dishes, etc., can read a very interesting article by F. A. Fuhrman (Fuhrman, 1986), where you can find a lot of scientific and semi-scientific facts.

Living Electricity Around Us

We have discussed why potential difference on the membrane of living cells appear, and have examined the process of the propagation of impulses along a nerve fiber. All the electric phenomena we are speaking about take place only on the cell membrane. But what was it that E. du Bois-Reymond registered in 1843 with the help of a simple galvanometer connected to a nerve? Since microelectrodes were not used until a hundred years later, it means that his galvanometer registered an electric field in the solution surrounding the nerve.

Examining the cable properties of a fiber, for the sake of simplicity we considered the outer solution of an electrolyte to be equipotential. Indeed, the voltage drop in the outer solution should be hundreds of times less than that inside a fiber because of much larger dimensions of the outer conductor (solution). Nevertheless, under sufficient amplification, an electric field can always be detected around an excited cell or organ, especially when all the cells of an organ are excited simultaneously. Our heart is such an organ in which all the cells are excited almost simultaneously. Like all other internal organs, it is all surrounded with electroconductive medium (the blood resistivity is ~ 100 Ohm·cm). Thus, at each excitation, the heart surrounds itself with an electric field. Therefore, a cardiologist, by measuring and analyzing the potential differences between different points of the body that appear in systole (electrocardiogram), sees how the heart works. In 1924, a Dutch physician Willem Einthoven was awarded the Nobel Prize for the introduction of the cardiograph technique into the diagnostics of cardiac diseases.

People knew for a long time that fish can be a source of electric discharges. You can see an electric catfish on ancient Egyptian tombs, and "electrotherapy" with the help of this fish was recommended by an ancient Greek physician Galen (130–200 A.D.). Another ancient doctor who treated Roman emperor Claudius (first century A.D.) prescribed electrical treatment in the following way: "A headache, even if it is chronic and unbearable, vanishes, if a live black ray is placed on a painful spot and is kept there until the pain disappears." Gout was treated in the same way: "With any type of gout, when pain starts, a live black ray should be placed under the feet. Meanwhile the patient should stand on wet sand washed by seawater and remain so until his leg below the knee goes numb." At the same time people noticed that the shock of a ray could pass through iron spears or sticks moistened with seawater and thus affect people who have no immediate contact with the ray.

Some fish are able to produce very strong discharges, immobilizing (paralyzing) other fish and even animals of a human size. Ancient Greeks, who believed that an electric ray could "enchant" both fish and fishermen, called it *narke*. This Greek word means a *making rigid*, or *striking* fish. The word *narcotic* is of the same origin.

Before the electric theory appeared, the theory that explained the shock of a ray as a mechanical effect was considered most efficient. Among the protagonists of this theory was French natural scientist R. Reaumur, whose name was given to one of the temperature scales. Reaumur assumed that a ray produces a shock by a muscle that can contract at a high rate. This is why if you touch such a muscle, a limb can go numb for some time, as it happens, for example, when you hit your elbow.

It was not until the end of the eighteenth century that the experiments which revealed the electric nature of the shock produced by a ray were carried out. A Leyden jar, the main electric capacitor of the time, played a certain part. Those who experienced the discharges of the Leyden jar and a ray claimed that the two were very similar in their effect. Like the discharge of a Leyden jar, the shock of a ray can simultaneously affect several people holding one another by hand, as long as one of the hands touches the ray.

The last doubts regarding the nature of the shock produced by a ray disappeared in 1776, when it was demonstrated that under certain circumstances, this shock could bring about an electric spark. For this purpose two metal wires with the tiniest possible air gap between them were partially submerged into a tank where the fish swam. A brief closure of the wires attracted the fish's attention, and it moved close to the wires and produced an electric shock; sometimes at this very moment sparks flashed between the wires. For the sparks to be more visible, the experiments were carried out at night. Soon after the experiments were over, some London newspapers started advertising a shake-up with a discharge of an electric fish for only two shillings and six pence. Benjamin Franklin, one of the authors of the theory of electricity, supported electric treatment. That is why the use of static electricity in medicine is still called *franklinization.*

By the early nineteenth century people already had known that the discharge of electric fish can pass through metals, but not through glass and air. In the eighteenth and nineteenth centuries physicists often used electric fish as a source of electric current. For example, Faraday showed while studying the discharges of an electric ray that the animal electricity did not dramatically differ from other kinds of electricity, which then were considered to be five: static (produced by rubbing), thermal, magnetic, chemical, and animal. Faraday believed that if people understood the nature of animal electricity, it would be possible "to convert the electric force into the nervous."

The strongest discharges are produced by the South American electric eel. They can be as strong as 500–600 V. The impulses of an electric ray can have voltage of up to 50 V and discharge current more than 10A so that their power often is more than 500 W. All the fish that produce electric discharges use special electric organs. In high-voltage electric fish, such as marine electric ray, freshwater electric eel, and catfish, these organs may take up a considerable part of the animal body volume. For example, in an electric eel they are located along the whole body, which is about 40% of the total volume of the fish.

The diagram of an electric organ is shown in Figure 1.11. It consists of electrocytes — much flattened cells packed into stacks. The endings of nerve fibers reach up to the membrane of one of the two flat sides of an electrocyte

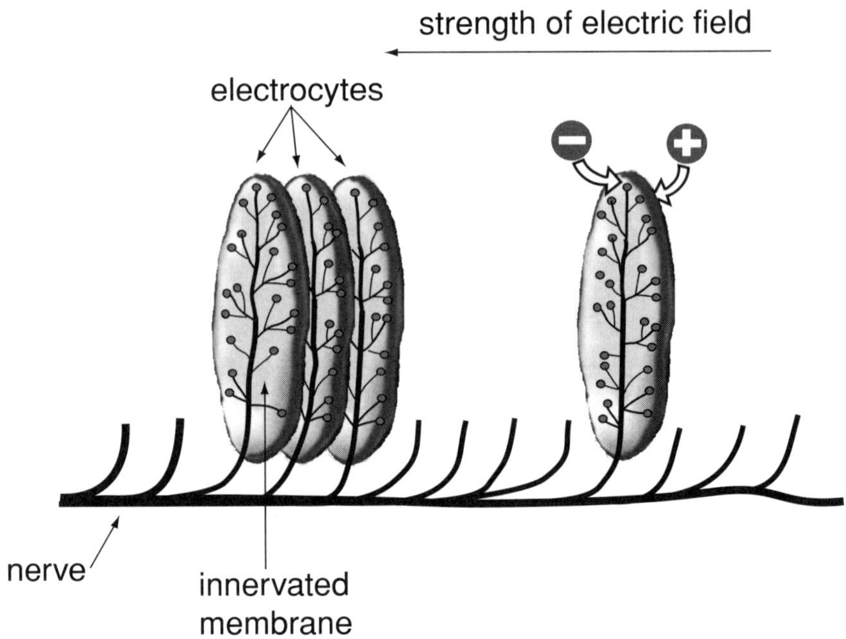

FIGURE 1.11. Diagram of an electric organ of a fish.

(innervated membrane); there are no endings on the other side (non-innervated membrane). The electrocytes are collected in a stack so that they face one another with opposite membranes. At rest, the potential difference at both membranes of an electrocyte is the same (about 80 mV); the inner medium of a cell bears a negative charge with respect to the outer. Consequently, there is no potential difference between the outer surfaces of both membranes of the electrocyte.

When an impulse approaches the electrocyte along a nerve (such impulses arrive at all electrocytes of an organ almost simultaneously), the nerve endings secrete acetylcholine. Acting upon the innervated membrane of the electrocyte, it increases the permeability of the membrane for sodium ions as well as some others; this results in the excitation of the membrane. During the excitation, the voltage across the innervated membrane of an electrocyte changes its sign and reaches $\sim +70\,\text{mV}$, whereas the potential difference between the outer surfaces of the same electrocyte becomes $\sim 150\,\text{mV}$. Since electrocytes are placed in a stack, the voltage between the end cells in the stack will be proportional to their number.

The number of electrocytes in one stack of the electric organ of an electric eel can be as many as five to ten thousand, which explains the high discharge voltage in these species. The magnitude of the discharge current is determined by the number of such stacks in the electric organ. There are 45 stacks of this kind on every fin of an electric ray, and an electric eel has about 70 of them on every side of the body. To prevent the current generated by the electric organ from passing through the fish itself, the organ is covered by insulating tissue with high resistivity, and it contacts outer medium only.

Electrical Compass

Interacting with the earth's magnetic field, various kinds of ocean and sea currents generate electric fields. Being highly sensitive to external electric fields, sharks and rays use them to find their way drifting along with these currents over boundless oceans. When the animals move with respect to the water, their motion is accompanied by the appearance of an electric field that serves as a kind of electromagnetic compass.

Is it really so? Let us see what happens in the coordinate system that swims along with a shark through motionless water. If the shark is assumed to be moving rectilinearly and uniformly the reference system we select will be inertial. It means that according to the principle of independence (equivalence) it does not matter in what coordinate system we consider the forces acting upon the shark: the resting (R) one or the one that swims together with the shark (S). However, in system S, the shark is not moving, so this motionless shark would seem incapable of establishing an electric field. Can it be the very case when the principle of equivalence is violated?

Of course not. Let's consider the motion of charged bodies in magnetic field in detail. It is well known that a charge e moving at the velocity V in the magnetic field with magnetic flux density B, is acted upon by Lorentz' force

$$F = e[VB]$$

In other words, a charge moving in a magnetic field establishes an electric field with strength $E = [VB]$. However, if we consider the coordinate system moving together with the charge, the Lorentz' force will not act upon the charge, since in the new system it is at rest. Therefore, we find it necessary to postulate the existence of an electric field with the strength $[VB]$ in a new coordinate system.

Hence, when passing over from an inertial coordinate system to another system moving with respect to the first one at velocity V, we should change the value of the electric field strength by $[VB]$.

Now, let us return to our sharks. Let one of them swim at the horizontal velocity V relative to motionless water in the earth-bound coordinate system. Now it is obvious to us that the interaction with the earth's magnetic field generates an electric field. The vertical component of the latter is equal to $[VB_{\mathrm{h}}]$ where B_{h} is the vector of the horizontal component of the magnetic field.

Thus, a shark swimming in the earth's magnetic field surrounds itself with an electric field and is a source of the electromotive force (Figure 1.12). Measuring the strength of the field, the shark can obviously evaluate its speed with respect to the earth.

Cartilaginous fish (like a shark) are very sensitive to electric fields, and their reaction is demonstrated in the relation to the gradients of the potential below 5 nV/cm in the frequency range from 0 to 8 Hz. A number of special experiments were carried out where nocturnal ocean predators were led to an electrical model of a prey which established an electric field with the gradient of 5 nV/cm. When it was accompanied by the smell of a typical prey, the sharks hit the model quickly and accurately from a distance of up to 0.5 m.

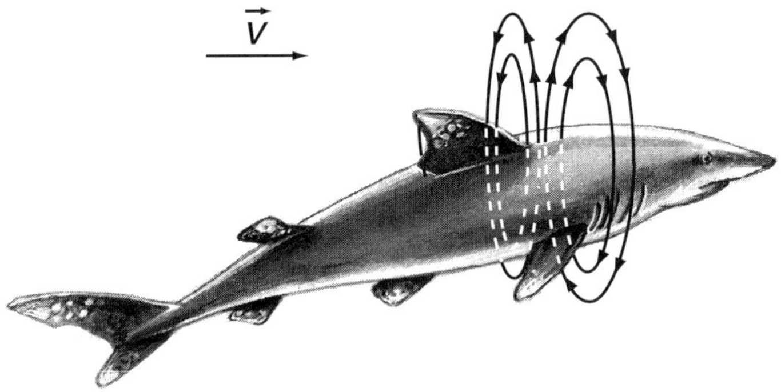

FIGURE 1.12. Shark that is swimming eastward at a horizontal velocity V. Shown around the shark are the electric field lines.

High sensitivity of cartilaginous fish to the electric field is explained by the fact that they have special electroreceptors, the ampullae of Lorenzini, each of which consists of a cross-sectionally small, but sufficiently long pore that is filled with jelly-like matter and ends with the receptor cells. The sensitivity of this live voltmeter was increased by the walls of the pore leading to receptor cells being made of a material with a much higher resistivity than the jelly-like matter.

Among the electric fish there are those that use their electric organ to look for the prey rather than for attack or defense. These are sharks, lampreys, and some catfish having very high sensitivity to an external electric field. A shark swimming in the open sea is known to be capable of finding sand-covered flatfish, an ability produced exclusively by the perception of bioelectrical potentials that appear when the prey is breathing. The electric organ of fish with the high sensitivity to an external electric field operates at a frequency of several hundred Hz and can generate oscillations of potential difference (about several Volts strong) on the surface of the animal body. As a result, an electric field appears which is picked up by electroreceptors, the ampullae of Lorenzini, the latter sending nerve impulses to the animal's brain. Since the conductivity of the objects in the water around the fish differs from that of the water itself, the electric field is distorted (Figure 1.13). The field distortions can be used by fish to take bearings in turbid water and to find quarry (von der Emde *et al.*, 1998).

It is fairly interesting to note that almost all fish that use their electric organs to find their way swimming keep their tails practically motionless. Unlike other fish, they move in water due only to undulatory movements of their well-developed lateral (electric ray) or dorsal (Nile pike) fins. Since the electric organs of these fish are located in the tail part of the body and the electroreceptors are in the middle, the strength of the electric field around of the receptors depends only on the conductivity of outer solution.

Japanese researchers established that immediately before a strong earthquake the catfish becomes unusually sensitive to weak mechanical disturbances, provided the aquarium is linked by channels with a natural water reservoir. This can be attributed to the potential differences that appear between points of the earth's crust before an earthquake and are perceived by the catfish. The strength of electric fields frequently emerging eight hours before the earthquake can reach 0.3 mV/m, which exceeds the sensitivity level of the fish more than a hundred times.

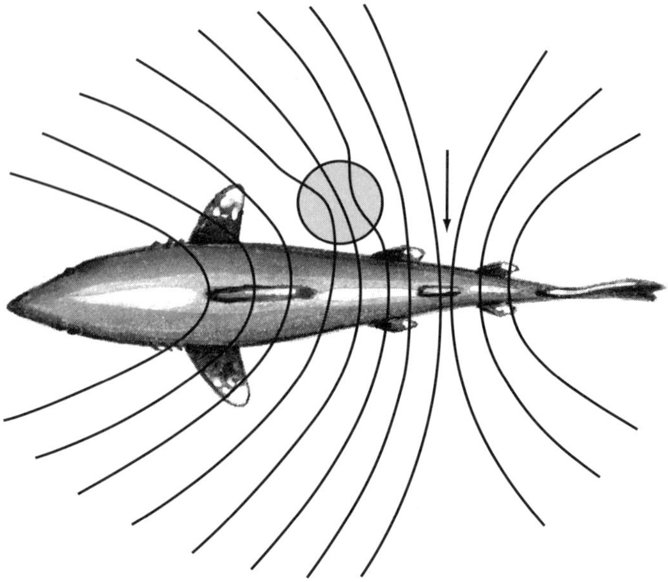

FIGURE 1.13. Distribution of equipotential lines of the electric field around a fish with an electric organ (its position in the tail part is shown by an arrow). The shaded ball-shaped object has a greater conductivity than the ambient one. It is clear that the electric field strength near the fish's lateral surface on the side of the object is different from that on the opposite side.

Interestingly, about two thousand years ago a legend was circulating in Japan, which said that the catfish was able to go underground, and by moving, cause earthquakes. However, it was not until the twentieth century that the behavior of animals before an earthquake attracted the attention of Japanese seismologists. Today, biological methods for predicting earthquakes are rapidly being developed.

It is well known that the fish in an aquarium with a direct current running through it swim in the direction of the anode and suddenly stop short of it, paralyzed. The voltage drop across the fish length must be about 0.4 V. After the current is turned off, the fish may "come to their senses" and start swimming again. If the voltage drop is increased to 2 V, the fish become rigid and soon die. The attractive force of the anode is successfully employed in electrical fishing. At the same time, the electric current scares away the fish that is more sensitive to it (e.g., sharks). Scientists carried out a series of experiments with a so-called electric fence to check its effect on sharks. It has been established that the current running between two electrodes acts as a barrier for sharks and is practically imperceptible for a nearby human.

Electricity in Plants

Plants are fastened in the soil by their roots and, therefore, often are considered a model of immobility. The notion is not entirely correct since all plants are capable of slow growth twists necessary to adapt to lighting and the direction of the force of gravity. Such movements are due to the nonidentical growth rate of different sides of an organ. In addition, plants move periodically during the day folding and unfolding their leaves and flower petals. Some other plants are capable of even more noticeable motion and respond with rapid movements to a variety of external factors: light, chemical agents, touch, vibration. Such sensitivity helped mimosa become proverbial: you only have to touch it slightly for its fine leaves to fold and for the main stem to droop. Various insectivorous plants and tendrils of lianas are also capable of fast response. How can such rapid movements occur in plants? What is crucially important here are the electric processes taking place in cells. A cell of a plant as well as a nerve or muscle cell of an animal has proved to have a potential difference of about –100 mV between the inner and outer surfaces of a membrane. The potential difference is affected by the different ion composition of intra- and extra-cellular media, as well as the nonidentical permeability of the membrane for the ions.

Under exposure to the external irritators listed earlier, the membrane of a plant cell is excited: its permeability for one of the cations (as a rule, for calcium) grows. As a result, the voltage across the membrane goes down to almost zero, but soon recovers its initial value. The duration of this action potential can be as long as 20–30 seconds (Figure 1.14), and can propagate from one cell to another in the same way a nerve impulse does, but at a noticeably slower speed. For example, the action potential propagates along a mimosa stem at a velocity of about 2 cm/s and along a leaf of insectivorous plant Venus flytrap at 10 cm/s.

The recovery of the initial potential difference across the membrane of a plant cell after its excitation occurs due to the opening of additional potassium membrane channels that were closed in the state of rest. The increase of potassium permeability of the membrane leads to the outflow of

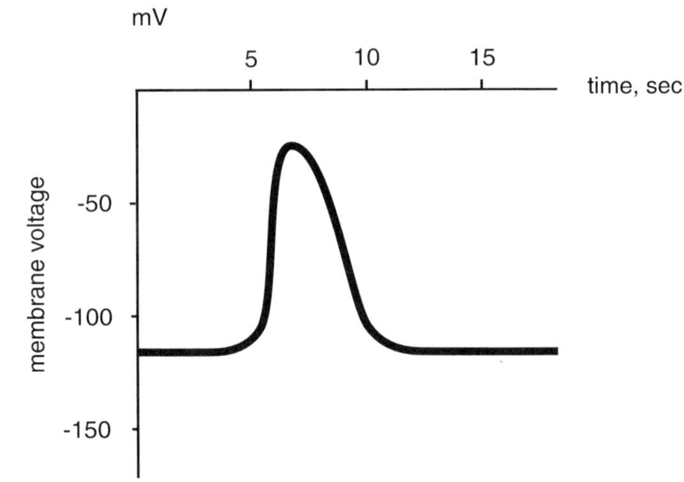

FIGURE 1.14. Action potential of a plant cell.

a certain number of potassium ions from the cell (there is more potassium inside than outside) and the recovery of the potential difference. The outflow of potassium ions from a plant cell under excitation is believed to happen not only due to the increase of the potassium permeability of its membrane, but also to other reasons not well studied. Thus, every excitation of a plant cell is accompanied by a short decrease of the concentration of potassium ions inside the cell and its increase outside, which underlies the motional response of the cell.

To understand the implications of the changing concentration of ions inside a plant cell, we can carry out the following experiment. Take some common cooking salt and put it into a small bag impermeable for the salt but permeable for water (e.g., cellophane). Place the bag with salt into a saucepan with water. You will soon see the bag swell, because the water would infiltrate the bag in order to balance the osmotic pressures inside and outside of the bag, both proportional to the concentration of dissolved ions. As a result, the growing hydrostatic pressure inside the bag may cause it to burst.

The living vegetable cells are concentrated solutions of salts, enclosed within a membrane, highly permeable for water. Once in contact with common water, the cells swell so much that the pressure inside them could rise up to $5 \cdot 10^6$ Pa. The magnitude of the intracellular pressure and the

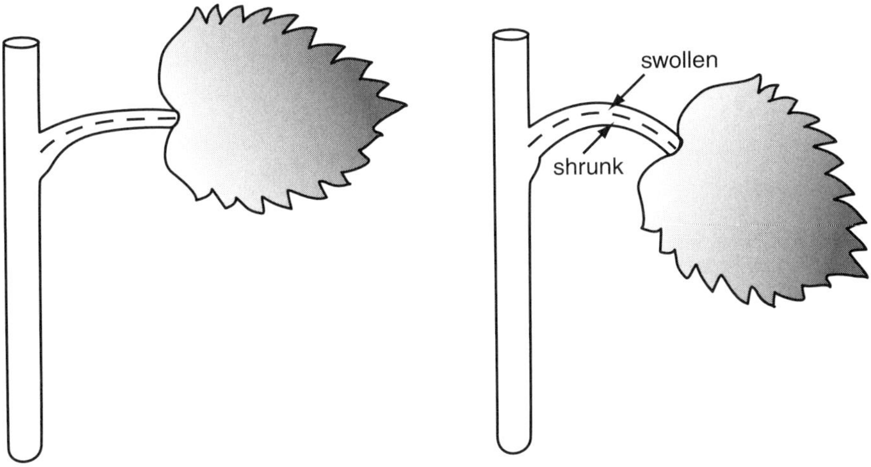

FIGURE Figure 1.15. The mechanism of motional activity of plants. *Left*: leaf stem in the state of rest; *right*: leaf stem under excitation.

extent of the plant cell swelling depend on the concentration of ions dissolved in it. For this reason, the decrease of the concentration of potassium ions inside a cell during excitation is accompanied by the lowering of intracellular pressure. Now imagine that some leaf stem consists of two cell groups arranged lengthwise (in Figure 1.15 the cell groups are separated with a dashed line) and the excitation spreads to the lower cell group only. Under excitation, the lower part of the stem collapses in, while its swollen upper part bends the stem. The same mechanism can work in other parts of a plant. Thereby, as is the case with animals, the electric signals that propagate along the plant provide an important means of communication between different cells, thus coordinating their activity.

2

Heart Pulse

In primitive society, as a man was cutting animal carcasses, he must have paid attention to the muscular sack in the middle of the animal chest. The "sack" could contract rhythmically for several minutes in the body of a listless prey. An apparent simultaneity of death and a heart arrest must have resulted in the theory that a man's heart is identified with his soul. That is why one of the earliest literary texts, *Odyssey,* by the Greek epic poet Homer, contains such word combinations as "grieve with one's heart," "fill the heart with courage," "heart's will," and so on. Another great Greek, Aristotle (384–322 B.C.), the founder of modern science, believed that the heart was the seat of thought.

The studies of the function of the heart were not to begin until much later, though. It was only in 1628 that the English physician William Harvey (1578–1657) published his famous *De motu cordis*, in which he expressed the conclusion that the heart served as a pump pushing the blood through blood vessels. To prove that blood circulated (flowed in a circle), Harvey merely had to show that a large quantity of blood passed out of the heart—roughly between 10 and 41 pounds of blood in half an hour—and that this quantity is greater than that which can be drawn from the body, or which can be obtained from processing food and liquid. He thus disposed of the

Galenical tradition according to which blood was considered to be made *de novo* in the liver, and to flow and ebb in the arteries. Harvey arrived at a conclusion that the same blood repeatedly returns to the heart, the latter acting as a hydraulic pump. To model the heart function, rather than using a pump with an ordinary valve, Harvey chose a special pump utilized at the time to drain water from mines.

Harvey's discovery provoked a long and heated discussion. Even Rene Descartes, an outstanding French physicist and mathematician, while agreeing with Harvey's blood circulation theory, did not share his opinion on the role of the heart in this process. Descartes considered the heart to be something that could nowadays be compared to a steam engine or even to an internal combustion engine. He considered the heart to be the source of heat that warmed the blood as it passed through this organ, thereby keeping up the heat in the whole body. In his opinion, this warmth was concentrated in the heart walls from the very beginning of the life. Thereby, when blood entered the heart cavity, it boiled up in it at once and further on, in the form of steam, got into a lung, which was constantly cooled by the incoming air. In the lung, the steam "condensed and turned back into blood." This is only a fraction of the fascinating story about the changes of people's views on the role of the heart in our organism (for review, see Leake, 1962).

Arteries, Blood, and Erythrocytes

Now we know that our heart is a pump operating in the impulse mode at a frequency of about 1 Hz. During each impulse, lasting for approximately 0.25 seconds, the heart of an adult is able to push about 0.1 L of blood from itself into the aorta.

The blood moves from the aorta into the narrower vessels, the *arteries*, which deliver the blood to the periphery. The origin of the word *artery* is fairly interesting—it derives from the Greek word *windpipe*. It is known that the greater part of a dead animal's blood is in the veins, the vessels through which the blood returns to the heart. That is why corpses have swollen veins and collapsed arteries. If such an artery is incised, it immediately assumes a cylindrical shape as it is filled with air. This, apparently, is the reason why a blood vessel got such an odd name.

The blood is a suspension of various cells in the aqueous solution. The red cells (*erythrocytes*) constitute the major part of blood cells. They take up about 45% of its volume, and each cubic millimeter of blood contains approximately 5 million erythrocytes. The volume occupied by the rest of the blood cells does not exceed 1%.

The inside of erythrocytes holds *hemoglobin*, a complex of the protein *globin* with the organic group (*heme*), the latter containing an atom of iron. It is hemoglobin that gives the erythrocytes (and all the blood) their red color, whereas hemoglobin's ability to reversibly bond with oxygen insures the high oxygen capacity of blood[1]. A liter of blood, free of erythrocytes, can bind only 3 mL of oxygen (at the atmospheric pressure), whereas a liter of normal blood can bind 200 mL. This enables the blood to perform its main function of supplying the cells of an organism with oxygen.

Erythrocytes are very flexible concavo-concave disks (Figure 2.1) and consist of a very thin (7.5 nm) membrane and liquid contents, a near-saturated hemoglobin solution. Despite the fact that the diameter of erythrocytes is about 8 μm, they can pass unscathed through the capillaries of 3-μm diameter. During this passage they become markedly deformed, resembling a parachute canopy, or rolled up into a pipe (see Figure 2.1 (b)). As a result, the area of contact of the erythrocyte with the capillary wall increases (as compared to the movement of a nondeformed erythrocyte) and so does the gas exchange rate.

How can we explain the ability of erythrocytes to deform easily? A body of a spherical shape can be shown to have the minimal surface area for the given volume. It means that, if an erythrocyte were a sphere, the area of its membrane would grow under any deformation. Therefore, the flexibility of these spherical erythrocytes would be limited by the rigidity of their cell membrane. Since a normal erythrocyte is nonspherical, it can deform without concurrent changes in the membrane surface area, and can readily assume all kinds of shapes.

There is a blood disease, known as hereditary spherocytosis, whereby erythrocytes are of spherical shape with a diameter of about 6μm. The membrane of such erythrocytes in their movement through thin capillaries is

---

[1] One of those who studied the mechanism that enables the release of oxygen from blood into body tissues was Christian Bohr, the father of the well-known physicist Niels H. Bohr. The dependence of the oxygen capacity of blood on the hydrogen-ion concentration is still referred to as the Bohr effect.

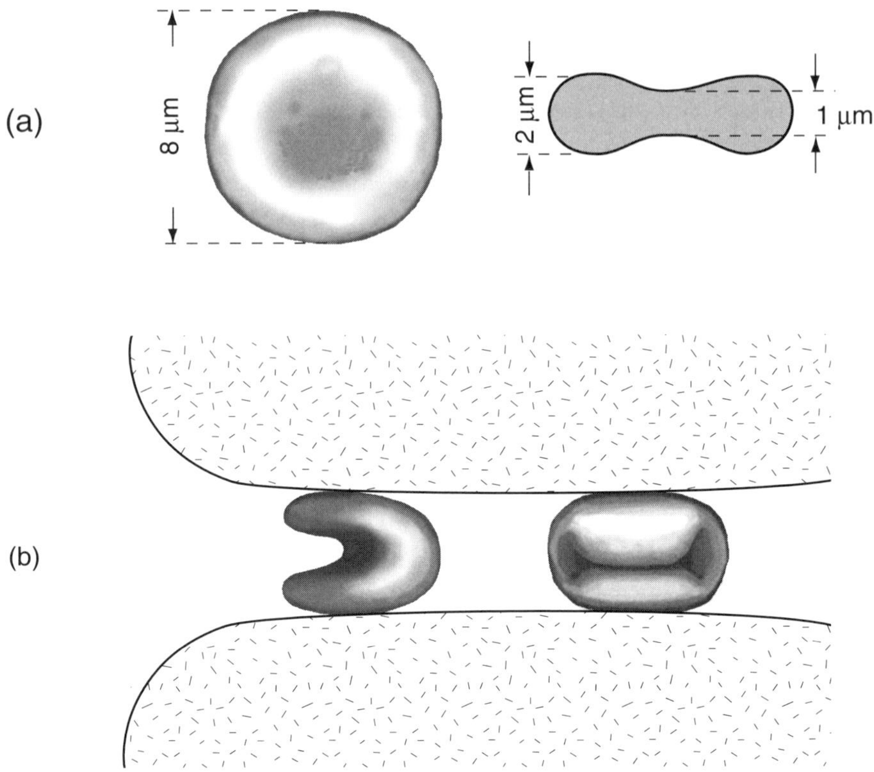

FIGURE 2.1. Erythrocyte: (a) undeformed (*left*: top view; *right*: side view); (b) assuming the shape of a pipe or parachute canopy during the passage through fine capillaries.

constantly in a state of stress and often ruptures. As a result the number of erythrocytes in the blood of these patients is lowered, and they suffer from anemia. A great number of people become deaf with age. Although only about 12% of the total adult population experience this gradual advent of deafness, its true cause could not be identified for a long time. It was only in the 1980s that scientists' attention was drawn to unusual properties of erythrocytes in deaf people. This conclusion was reached while measuring the viscosity of blood in normal and deaf patients with rotational viscosimeters, which make it possible to examine the dependence of the viscosity of blood on a velocity gradient (rate of shear).

The viscosity of blood, like that of most suspensions and solutions (in contrast to pure solvents) is dependent on the rate of shear (Figure 2.2). For this reason, the blood is termed as non-Newtonian liquid. This dependence is due to the fact that not only the blood plasma, but also erythrocytes with their relative orientation and shape and changeable during the motion, participate in the transfer of shear stress from layer to layer along a perpendicular to the flow velocity.

Since erythrocytes are not spherical, they may become oriented when the suspension is sheared. If the flat surface of the erythrocyte disc is parallel to the vessel wall, there will be a minimum disturbance to the lines of flow, and the presence of the cell will have the least effect on the viscosity of the system. Direct microscopic observation of suspension of erythrocytes, flowing through a tube, has shown that with increase in the rate of flow, the cells become increasingly oriented along the axis of the tube. Figure 2.3 shows schematically how the increase in shear stress can bring about the orientation of erythrocytes in the moving blood flow (a) and their deformation (b). It is probable, therefore, that orientation effects contribute

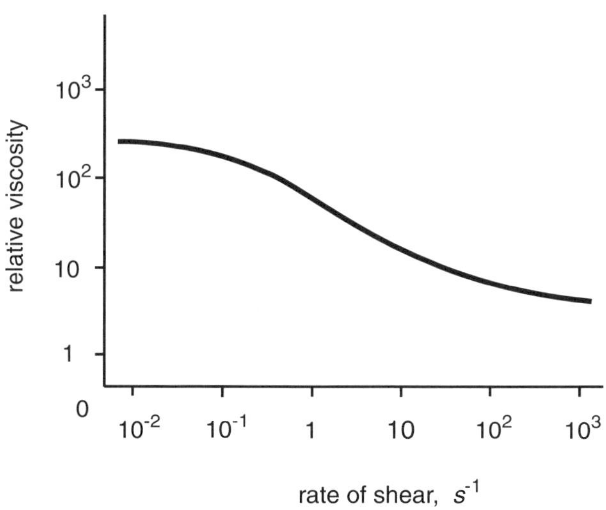

FIGURE 2.2. Dependence of relative viscosity of human blood on rate of shear. (From Gabelnick, H.L., and M. Litt (Eds.) *Rheology of Biological Systems*, 1973. Courtesy of Charles C. Thomas, Publisher, Ltd.: Springfield, IL.)

(a)

(b)

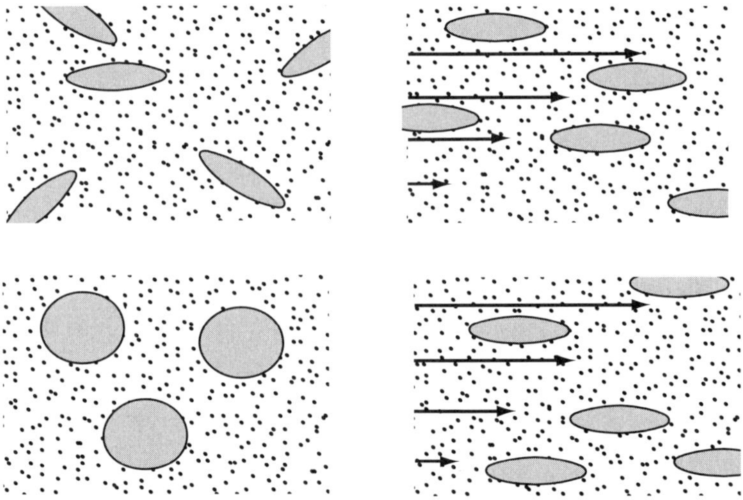

FIGURE 2.3. Schematic presentation of orientation of ellipsoidal (a) and deformation of spherical (b) cells in the flow. *Left*: stationary liquid; *right*: the same cells in a moving liquid with transverse velocity gradient. The vector field of velocities is schematically shown with arrows.

to the reduction in apparent viscosity of blood with increase in the rate of shear (Figure 2.2).

By comparing the viscosity values at very small and large values of rate of shear, it is possible to calculate certain deformation characteristics of erythrocytes. It turns out that the rigidity of erythrocytes in deaf people is much higher than in healthy ones. High flexibility (small rigidity) of erythrocytes is known to be a necessary condition of their penetration into the finest capillaries. Hair cells of cochlea, which perceive sonic vibrations and convert them into a sequence of nerve impulses, are supplied with blood through capillaries with a diameter of 4–5 μm. At the same time the diameter of the disk-shaped erythrocytes amounts to 7–8 μm. Therefore, it is only the high elasticity of erythrocytes (their ability to roll themselves up into a pipe) which helps them literally push through such narrow capillaries near hair cells.

Several different experimental approaches have been developed to assess red cell deformability. In some techniques, the transit times of individual erythrocytes across a membrane containing cylindrical micropores are measured electrically (Frank and Hochmuth, 1988). In filtration tests, flow rates of red cell suspensions through membranes are determined (Nash, 1990). Use of micropore transit to estimate red cell deformability has the advantage that the pore diameters can be chosen to correspond to the diameters of the smallest microvessels ranging from 3 to 7 μm.

The complexity of the deformations undergone by red cells makes it difficult to use a mathematical model to simulate the cells' transit across a membrane using a mathematical model. However, recently Secomb and Hsu (1996) described a model that enables the analysis of the red blood cell motion through cylindrical micropores. According to the model, both the transit time and the blood filterability are sensitive to the erythrocyte membrane shear viscosity. However, it should be noted that an axisymmetric model proposed by the authors may prove inadequate for actual cell passage across a pore.

It is believed that in deaf persons the capillaries near hair cells are clogged with stuck erythrocytes, unable to roll up into a pipe. As a result, these cells gradually die because of blood supply failure, so that deafness occurs. A sudden-onset deafness is often of the same nature as that developing over time in middle-aged people—stuck erythrocytes in fine capillaries of cochlea. As was found by Hall *et al.* (1991), the filterability of the red blood cells is significantly impaired in patients with the sudden deafness, which suggests that alterations in the cochlear microcirculation are linked to erythrocyte resistance to transient deformation.

However, sudden deafness may be unrelated to the change in the erythrocyte rigidity and, therefore, sometimes yields to efficient treatment. The blood of a patient is temporarily thinned in order to lower the concentration of erythrocytes in it. As a result, the motionless line of erythrocytes in the finest capillaries of cochlea is resolved and the patient regains his or her hearing.

One more practiced technique to estimate erythrocyte deformation is the electric field method presented by Engelhardt and Sackman (1988). A high-frequency electric field polarizes cells suspended in a low ionic strength solution in a frequency range where the Maxwell-Wagner polarization (Pohl, 1978) is effective (0.1–10 MHz), and thus generates within this frequency range a constant uniaxial force in the direction of the electric field strength. The cells respond to that force by elongation, which is mainly elastic shear deformation (deformation under constant volume and surface area).

Velocity of Pulse Wave

The blood motion across the vessels is a rather complicated process. The wall of the aorta, as well as the walls of all other arteries, is highly elastic: its Young's modulus is 10^5times smaller than that of metals. Therefore, when blood enters the aorta, the latter starts expanding until the blood inflow stops. After this, the elasticity of the expanded aorta wall, seeking to bring the aorta back to its initial dimensions, squeezes the blood out into the artery section, more distant from the heart (the reverse flow is hindered by the aortic valve of the heart). This section of the artery becomes stretched and everything is repeated from the beginning.

Were the deformation of an artery wall to be recorded simultaneously at two nonequal distances from the heart, it would turn out that the artery deformation reached its maximum values at different moments of time. The more distant from the heart, the later the vessel deformation reaches its peak value.

Thus, after each systole a wave of deformation runs along an artery in the direction from the heart to the periphery, in much the same way as waves propagate along a tight string or along the water surface from a stone dropped into it. And if a finger is placed on an artery, which is close to the body surface (e.g., near a wrist), the finger will sense the waves in the form of pushes (pulse).

Here, we should note that the velocity of propagation of the blood vessel deformation wave may significantly differ from that of the compression wave in blood. The latter is, obviously, equal to the velocity of propagation of sound and amounts to several hundred meters per second, whereas the deformation waves cover no more than several meters a second.

People had learned to measure frequency, rhythm, and repletion (amplitude) of pulse long before its origin became known. The first references to the measurements of pulse in humans date back to the third millenium B.C., when physicians used the registration of pulse for diagnostic purposes. The teaching on pulse became one of the major accomplishments of diagnostics in ancient China. Their concept, elaborated in the course of many years of cardiac pulse observations, was that all processes in an organism should somehow affect the shape of a pulse wave. They studied pulse at several points rather than at a single point on a wrist (as we do now). Sometimes there were as many as ten such points. Chinese physicians drew distinctions between 28 types of pulse: superficial, deep, rare, frequent, thin, excessive, free, tough, strained, and so on. Thousands of years before the scientific validation of the blood circulation theory, they had already assumed that it was the pulse that determined the "blood circle." The ease of pulse measurement (no instruments are needed except for a stopwatch) makes it one of the main indicators of one's state of health even at present.

The propagating wave of deformation of the artery walls got to be known as a pulse wave. Measuring the velocity of pulse wave propagation had met no success untill the early twentieth century, when the first inertialess recording devices appeared. The value of the velocity lies, as a rule, between 5 to 10 m/s and more, which is 10 times greater than the average speed of the movement of blood along blood vessels. The velocity of propagation of a pulse wave has proved to be dependent on the elasticity of an artery wall and, therefore, could be an indicator of its state in the diagnosis of various diseases.

A mathematical expression for velocity θ of the propagation of pulse waves in arteries is easily derived if we consider wave propagation in an infinitely long tube, the walls of which are completely elastic; that is to say, the wall material does not show any elastic hysteresis or internal friction, and its stress-strain relationship obeys the law of Hooke. Following is the well-known *Moens-Korteweg equation*:

$$\theta^2 = Ea/\rho d \tag{2.1}$$

where E is the Young's modulus of elasticity for artery wall material; a is the artery wall thickness; ρ is the density of blood; and d is the artery diameter.

The substitution of $a/d = 0.1$, $\mathrm{E} = 10^6\ \mathrm{N/m^2}$ and $\rho = 10^3\ \mathrm{kg/m^3}$ into Eq. (2.1) yields the value $\theta \approx 10$ m/s, which is close to the average value of the velocity of propagation of pulse wave, measured experimentally. Anatomical studies show the quantity *a/d* to vary a little from one person to another and to be practically independent of the artery type. Therefore, in view of the constancy of *a/d*, the pulse wave velocity can be considered to change solely with changes in the elasticity of an artery wall, its Young's modulus.

With age, as well as in the case of diseases accompanied by increased E of artery walls (hypertension, atherosclerosis), E may grow almost twofold compared to the norm. This makes it possible to use the measurement for diagnostic purposes.

It is interesting to note that Eq. (2.1) for the velocity of propagation of the pulse wave in the arteries was first derived by the famous English scholar Thomas Young in 1809. Young is remembered mainly as the creator of the wave theory of light, and also because the elasticity modulus of materials is named after him. He was also the author of the classical works on blood circulation theory, including those on propagation of pulse wave in arteries. He was truly an extraordinary personality. He could read at the age of two, and at 14 had a good command of ten languages, played nearly every musical instrument, and had the skills of a circus performer. During his entire life he combined two professions, that of a practicing physician and of a physicist.

Reflection of Pulse Waves

Like any waves, the pulse waves in arteries are capable of reflecting from the sites where the conditions of their propagation change. Such sites for the pulse waves are the regions of artery branching (Figure 2.4(a)). The wave reflected from the branching region is superimposed on the primary one, which makes the curve of blood pressure changes in a vessel bimodal (Figure 2.4(b)). It is possible to estimate the distance between the point of the recorded pressure and the site of branching by measuring the interval between two maxima on the pressure curve and the given velocity of propagation of a pulse wave. Sometimes the pressure curve exhibits more than two maxima, which is indicative of the multiplicity of the pulse wave reflection process.

The reflected pulse wave, like the primary one, is accompanied by the arterial wall deformation. However, whereas the elastic energy of the wall deformation caused by the propagation of the primary wave is later converted into the kinetic energy of the blood moving from the heart to the periphery, the reflected wave in fact obstructs a normal blood flow. For this

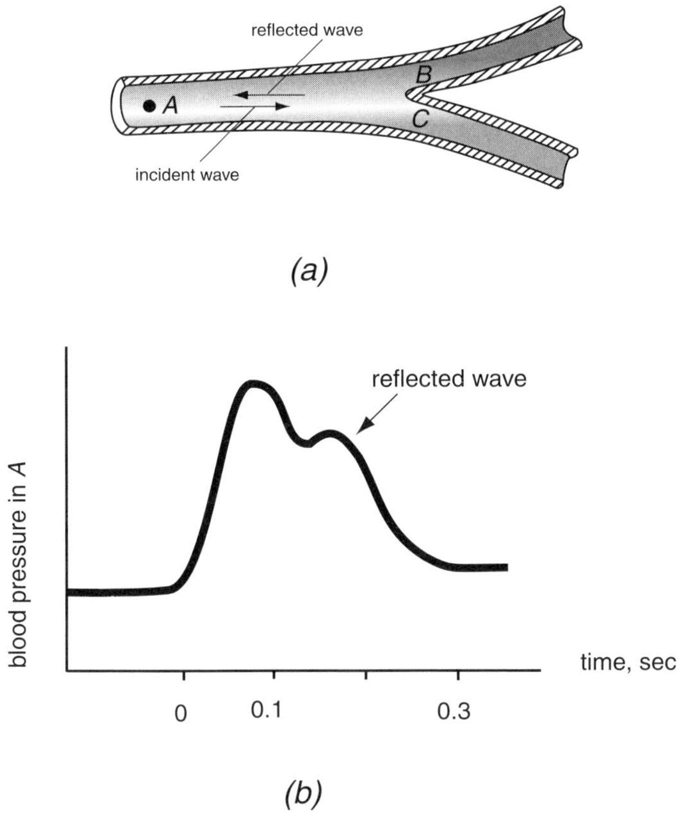

FIGURE 2.4. Emergence of reflected pulse wave at the site of artery branching. (a) section of branching artery; (b) arterial pressure change in the presence of a reflected wave.

reason, the pulse wave reflection interferes with the normal functioning of our circulatory system.

It can be shown (Hardung, 1962) that under certain assumptions, the amplitude, P_r, of pressure changes in the reflected pulse wave depends on the parameters of branching and the amplitude, P_i, of the primary (incident) pulse wave:

$$P_r(S_A/\theta_A + S_B/\theta_B + S_C/\theta_C) = P_i(S_A/\theta_A - S_B/\theta_B - S_C/\theta_C) \qquad (2.2)$$

where θ_A, θ_B, θ_C are the velocities of propagation of a pulse wave along arteries A, B, and C, and S_A, S_B, and S_C are the cross-section areas of the arteries. It can be concluded from Eq. (2.2) that there is no reflected wave if the factor in the parentheses in the right-hand part of the equation is zero. If the velocity of the pulse wave propagation is considered to be the same beyond the branching site since, as a rule, *a/d* and E remain unchanged, there will be no reflected wave provided:

$$S_A = S_B + S_C \tag{2.3}$$

Thus, in order to have no reflected waves at the sites of arterial branching, it is necessary that the sum total of the cross-section areas of arteries remain unchanged beyond the branching. In other words, to minimize the energy loss due to reflected waves in the circulatory system, shown schematically in Figure 2.5, the following rule should be observed:

$$R_0^2 = \text{SUM}\ (R_{iK}^2), \quad i = 1 \ldots N_K \tag{2.4}$$

where R_0 is the aorta radius and R_{iK} is the radius of one of N_K vessels, which are K branchings away from the aorta.

Note that the greater part of branchings of major arteries satisfies, to

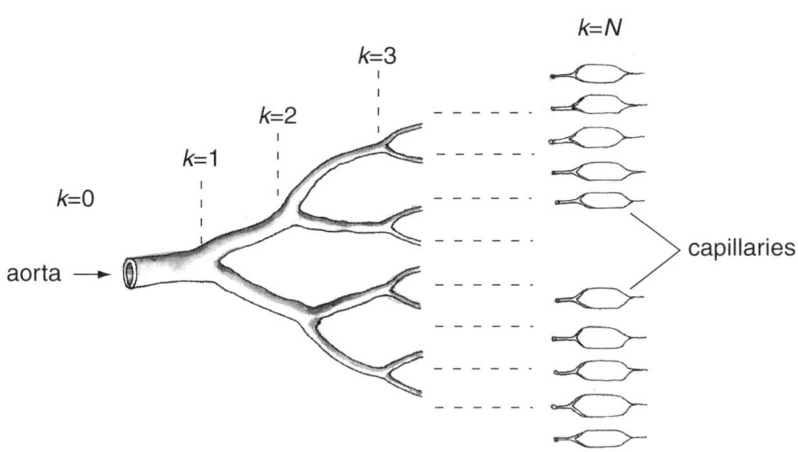

FIGURE 2.5. Scheme of arterial bed. Branching number $k = 0$ corresponds to the aorta; $k = N$ corresponds to capillaries.

some extent, Eqs. (2.3) and (2.4), which requires the constancy of the cross-section of the blood vessel bed before and after the branching site. In some cases, the equation does not hold, and it leads to this.

Equilibrium of the Blood Vessel Wall: Aneurysm

After each systole the blood pressure in the aorta increases, its walls stretch, and a pulse wave propagates along them. This rhythmic extension recurs about 100,000 times a day and approximately 2.5 billion times throughout life. In principle, the arterial wall structure is capable of withstanding these rhythmic hydraulic shocks. Sometimes, however, the aorta wall gives way and starts expanding to form an *aneurysm*. Once started, the expansion tends to grow on and on, so that the aneurysm finally bursts, causing death. The probability of aneurysm occurrence increases with age.

The typical place for an aneurysm to develop is the abdominal part of the aorta, a little above its branching. The aneurysm is believed to appear in the region of pulse wave reflection from the site of aorta branching. As shown in Eq. (2.2), the amplitude of the reflected wave is proportional to the difference between the cross-section area of a vessel before the branching and the total cross-section area beyond it. The difference grows with age due to narrowing of the peripheral arteries. As a result, the amplitude of the reflected pulse wave increases, causing a greater extension of aorta walls in that place.

The growth of an aneurysm is a manifestation of the law of Laplace, describing the relationship between tension T, stretching the blood vessel wall (the ratio of force to the area of longitudinal section of the vessel wall), its radius R, excess pressure P inside the vessel (transmural pressure), and its wall thickness a (see Figure 2.6(a)):

$$T = PR/a \qquad (2.5)$$

Another statement of the law of Laplace is used more often, where the left-hand part involves the product Ta, which is numerically equal to the force stretching the vessel wall and applied to its unit length. In such cases, assuming $Ta = T'$, we have the following form of the law of Laplace:

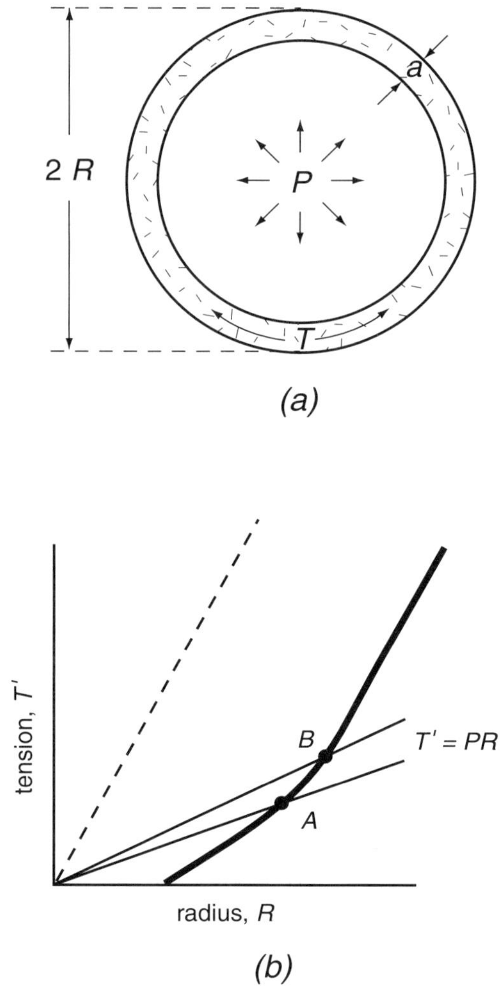

FIGURE 2.6. Equilibrium for normal blood vessel wall under elastic tension: (a) the forces that are in equilibrium at the blood vessel wall (shown in cross section); (b) estimation of change in vessel radius occurring due to change in blood pressure.

$$T' = PR \tag{2.6}$$

For equilibrium, then, the transmural pressure must always equal the circumferential tension divided by the radius. This means that the circumferential tension in the blood vessel wall has a sort of mechanical advantage in opposing the transmural pressure.

It follows from Eq. (2.5) that, with P increasing, T should grow, too, which causes the vessel wall to stretch and its radius R to become larger. However, since the aorta wall volume can be considered a constant, the aorta radius increase should be accompanied by the thinning of its wall. Therefore, with P increasing, ratio R/a should grow as well, causing still greater T, and so on. Thus, any increase of arterial pressure would seem to result in an avalanche-like growth of R and decrease of a, leading to formation and blowout of an aneurysm. Then why does this actually occur in very rare cases and, as a rule, at a middle age?

The aorta of a man has an inside diameter of about 2.5 cm and the wall thickness of 2 mm. The wall consists of cells containing two basic types of elastic materials, *elastin* and *collagen*. In an unstrained vessel wall, the collagen fibers are not straightened in full. So, the elasticity of an aorta wall under small deformations is determined by the easily stretched elastin. Under large deformations, the mechanical properties of an aorta wall are determined by collagen, which exhibits a much higher rigidity than elastin. Therefore, the dependence of tension T', which stretches aorta walls, on its radius displays a noticeable inflection and can be approximated by the segments of two straight lines with slopes that correspond to elasticity of elastin and collagen, respectively (Figure 2.6(b)).

Let's examine the stability of the equilibrium of a vessel under the transmural pressure. The curved line in Figure 2.6(b) represents the relation between elastic tension and stretch. For equilibrium we must have the law of Laplace (2.6), which is represented by a straight line through the origin, the slope of which (tangent of the angle) is equal to the transmural pressure, P. The point A, where the Laplacian line and the curve intersect, represents the point of equilibrium. If the pressure were increased, a Laplacian line of increased slope would intersect the curve at B, indicating the increased radius at the increased transmural pressure. The equilibrium under transmural pressure and elastic tension is completely stable if the pressure changes in the indicated range.

However, it must be recognized that if the transmural pressure is great enough, equilibrium may not be possible. In a certain case, as depicted in Figure 2.6(b) the curve of tension versus radius for arteries does not continue to increase in slope, but becomes a straight line. If the Laplacian line has a slope great enough to be parallel to this final slope of the elastic line (see a dotted line in Figure 2.6(b)), no intersection is possible. The vessel radius will increase until the vessel bursts. A similar process occurs under the blowout of aneurysm.

The cause of the appearance and blowout of an aneurysm lies not only in the increased amplitude of arterial pressure but also in changes of mechanical properties of the arterial wall. The properties of collagen change in the elderly. It becomes less rigid, so that the aorta wall becomes easily stretched.

In young people, the slope of the collagen part of the tension-radius curve corresponds to transmural pressure of about 130 kPa (1000 mm Hg), which is 6–8 times their actual arterial pressure. In middle-aged people, the rigidity of an aorta wall can decrease almost fivefold, and the arterial pressure can be as high as 26 kPa (200 mm Hg). This increases the probability of the emergence and the blowout of the aneurysm.

Murray's Law

It is believed that biological form is regulated by a set of physiological principles that optimize function. Therefore, to operate efficiently, the mammalian blood circulation must obviously comply with a number of physical design principles. One of the principles described earlier calls for the minimization of a pulse wave reflection and the observance of Eq. (2.4) at sites of vessel branching. However, as follows from Table 2.1, Eq. (2.4) does not hold for branching of fine vessels, and the total cross-section area of capillaries is almost 1000 times the aorta cross-section area. This can be explained by the fact that the pulse wave amplitude decreases as the wave propagates from the heart to periphery, and so the blood motion through capillaries is practically uniform. This is why there is no need to observe Eq. (2.4) in the capillary part of the blood bed.

There are, however, other principles, compliance with which helps to minimize expenses of energy for blood circulation. Murray (1926) also hypothesized that the design of the vascular system is such that the operating costs of the circulatory system are minimized. The operating costs consist of the cardiac work incurred in generating the pressure that drives the flow of blood and the metabolic work needed to make and maintain the blood.

Table 2.1 Arterial circulatory system of humans. For all four vessel types, the SUM (r^2), which should be constant by anti-reflection law, has been calculated (see Eq.z 2.4). Values for SUM (r^3) should be constant by Murray's law. Values for SUM (r^4) are given for comparison. Modified from LaBarbera (1990). (Reproduced by copyright permission of Science.)

Vessel	Average radius, cm	Number of vessels	*SUM* (r^2) "no reflection" law	*SUM* (r^3) Murray's law	*SUM* (r^4)
Aorta	1.25	1	1.6	1.95	2.44
Arteries	0.2	159	6.4	1.27	0.25
Arterioles	0.003	1.4×10^7	127.4	0.38	0.001
Capillaries	0.0004	3.9×10^9	1432	0.86	0.0005

According to Murray's law, the total power required to sustain a regulated flow of blood through a segment of a blood vessel is assumed to be

$$P = P_f + P_m \tag{2.7}$$

where P_f is the power required to drive the flow of blood, and P_m is the metabolic power required to maintain the blood supply.

For Hagen-Poiseuille flow, the power needed to overcome the viscous drag on the blood flow through the vascular segment of radius R, and length L is

$$P_f = (8\mu Q^2 L)/(\pi R^4) \tag{2.8}$$

The power required for maintaining the blood supply is assumed to be proportionate to the blood volume, giving

$$P_m = \alpha \pi R^2 L \tag{2.9}$$

where α is a metabolic constant for blood. We seek the optimal vessel geometry for given mean pressure and flow. Hence, after substitution of (2.8) and (2.9) into (2.7), the minimization conditions

$$\partial P/\partial R = 0 \quad \text{and} \quad \partial^2 P/\partial R^2 > 0$$

give

$$Q = R^3(\alpha/\mu)^{1/2}\pi/4 \tag{2.10}$$

Proceeding from similar considerations, Murray concluded that in an optimal vessel system, volume flow is proportionate to a vessel diameter, cubed. Assuming that total volume flow doesn't change at any branch point, Murray derived the following relation.

Murray's law: In an optimally designed system involving bulk laminar flow of a Newtonian fluid through pipes, at any branch point the radius of the parent vessel (R_0) cubed will equal the sum of the cubes of the radii of the daughter vessels ($R_1, R_2, R_3, \ldots, R_n$):

$$R_0^3 = R_1^3 + R_2^3 + \ldots + R_n^3 \tag{2.11}$$

One of the implications of Murray's law is that in the optimal circulation system the shear stress at the vessel wall, τ_w, is uniform throughout the system. Indeed, following from Eq. (2.10),

$$\tau_w = 4\mu Q/(\pi R^3) = (\alpha\mu)^{1/2} \tag{2.12}$$

As follows from Table 2.1, the variation of a radius of vessels along the blood bed is best described in terms of Murray's law, though even this law seems to be imperfect (cf. 1.95 for aorta with 0.38 for arterioles). Therefore, new attempts have been made to find a more adequate law to describe the branching of blood system vessels (for example see, Taber, 1998).

Blood Circulation in Giraffe and Space Medicine

Who would not dream about flying into space and seeing the earth from the outside? Unfortunately, this dream will come true for only a few of us, so difficult and still dangerous is the job of an astronaut. Thousands, even tens of thousands of people get the spaceship to flight and solve the flight-related problems. A significant part of such problems pertains to a new field in biology called *space biology*.

The first thing an astronaut faces at take-off is the acceleration, when the spaceship rapidly picks up speed. In the course of launching the spaceship into orbit as the earth's artificial satellite, for almost five minutes an astronaut is subjected to acceleration, its magnitude varying between g and $7g$. In other words, the astronaut's weight during the ship's blast-off may grow sevenfold.

Accelerations also affect an astronaut as the spaceship enters the dense layers of the atmosphere when returning to the earth. Naturally, the increase in the astronaut's weight hinders his or her movements. Just try and imagine how difficult it would be to raise a sevenfold heavier hand to operate a toggle-switch on the control panel. Therefore, at times of overload (during the blast-off and deceleration), most operations related to the control of the ship should be automated.

However, the difficulty of performing various movements due to the increased astronaut's weight is only one aspect of the effect of accelerations in a space flight, comparatively easy to bear. A greater danger lies in that accelerations also result in displacements of soft tissues and some internal organs along the direction of inertial forces.

The most mobile part of an organism is, obviously, the blood. Therefore, under exposure to accelerations, the most significant changes arise in the circulatory system. If the acceleration is directed from pelvis to head, the effect of the inertial forces brings about the deflux of blood from the vessels of the head and its afflux to organs of the torso's lower part. This may result in vision disturbances and even fainting. If the action of acceleration in that direction lasts for a minute, its maximum value should not exceed $3g$.

Were the blood vessel walls absolutely rigid, the effect of inertial forces would fail to cause the redistribution of blood in an organism. All effects of acceleration in a circulatory system are due to the fact that blood vessel walls are highly distensible; owing to this distensibility, the change of blood pressure can alter the volume of blood vessels and of the blood they contain.

The pressure of water in a vessel within the earth's gravity field increases with depth, so that 10 cm submergence corresponds to 1 kPa pressure growth. If a vessel moves with acceleration ng, the water pressure will be diminishing in the direction of this acceleration vector by n kPa every 10 cm.

Arterial blood pressure of a healthy man at the heart level is 16–18 kPa. In the sitting position, the head is approximately 40 cm above the heart level, so in the absence of accelerations the blood pressure in the large arteries of the head is equal to 12–14 kPa, which is quite enough for distension of the arteries.

During the motion with acceleration $3g$ in the direction from pelvis to head, the arterial pressure in the vessels of a head diminishes by another 12 kPa, and becomes practically equal to the atmospheric pressure. Blood vessels collapse, the blood flow through them reduces sharply. Therefore, under such accelerations, the brain cells begin to suffer from the lack of oxygen, which brings about the loss of consciousness.

For the same reasons, the pressure in the vessels of lower extremities under upward accelerations grows, and may reach 75 kPa at $3g$. A more than fourfold increase of the arterial pressure brings about the excessive expansion of the vessels. As a result, the volume of blood in lower parts of the body increases and in upper parts diminishes. Besides, the vessels of the lower part of the body under the action of enormous pressure start leaking water through their walls into surrounding tissues. This leads to swelling of legs, called *edema*.

How is normal blood circulation in an astronaut and a jet plane pilot to be ensured under the action of accelerations? The simplest solution is to position them so that their dimensions along the direction of acceleration vector were minimal. Then the arterial pressure in different parts of the body would vary insignificantly and no redistribution of blood would occur. This is why astronauts blast off and land in a reclining position.

Interestingly, the same conclusion (to fly "lying sideways") was arrived at by the characters of Jules Verne's novel *From the Earth to the Moon*, written in 1870. Apparently, this novel by the great French science-fiction writer can claim priority in touching upon the main problems of space medicine.

But what should the pilots of jet airplanes do? When executing a sharp maneuver, they cannot be in a lying position, as they have to control the craft at that very time. What if a pilot is dressed in a tight-fitting suit with water placed between its inside and outside layers? Then, under accelerations, the water pressure in any section of such a suit would change by the same value as the pressure in nearby blood vessels. Therefore, despite the pressure inside the vessel still growing, the latter would be unable to expand. No redistribution of blood would ensue. Such a suit became known as a *g-suit* and is successfully employed in astronautics and supersonic aviation.

The majority of animals inhabiting the Earth are horizontal—the brain and the heart, the two most important organs, are located inside their bodies at the same level. This is very convenient. No additional effort is required by the heart to supply the brain with blood.

Man cannot be termed as a horizontal animal, which is why he has relatively high arterial pressure. The hypertensives of the same kind include some birds (a rooster, for instance) and, of course, a giraffe.

The heart of typical horizontal animals is unable to maintain the blood supply of the brain in an unnatural posture. For example, were a rabbit or a snake put in vertical position, they would very soon lose consciousness because of brain anemia.

It turned out that an analog of a g-suit can be found in a giraffe. Of course, this does not imply that a giraffe is an alien from space. The necessity to wear such a suit on earth is explicable by an extraordinary great height of the animal: it may reach 5.5 m.

The giraffe's heart is at a height of about 2.5 m, so the blood vessels of legs should experience a colossal pressure of all this column of liquid. What is it that saves a giraffe's legs from developing edema? Between the vessels of a giraffe's legs and its dense skin is a lot of intercellular liquid, which saves the vessels from excessive expansion in much the same way as the water in a pilot's g-suit.

But how can giraffe's blood flow to the level of its brain, 3 m higher than his heart? If a giraffe had the heart blood pressure comparable to that of humans, then at the level of a head it would be lower than atmospheric, so that blood would not be able to pass through the brain. It is unsurprising, therefore, that a giraffe is a hypertensive. Its arterial pressure at the level of the heart may be as high as 50 kPa. This is the price a giraffe has to pay for its great height.

The present day's fashion makes young people wear an analog of a g-suit — tight jeans. Doctors maintain that tight-fitting pants can help victims of severe injuries below the belt avoid a sudden drop of arterial pressure, typical in blood loss.

How Blood Pressure and Blood Flow are Measured

One of the main indicators of the functioning of the heart is the pressure with which it delivers blood into vessels. In 1733, the Reverend Stephen Hales (1677–1761), a conventional vicar of Teddington, reported on his amazing direct measurement of blood pressure in a variety of animals, including horses. Of interest are the circumstances that brought Hales to the discovery of arterial pressure. Before taking up the research on the forces setting the blood of animals in motion, he had devoted several years to studying plants. In particular, he was interested in what makes the sap rise from the roots to the leaves of a tree. The results of his studies can be found in his book *Vegetable Staticks* published in 1727.

Proceeding from the assumption that vegetable sap in a tree has the same role as blood in an animal, Hales undertook the investigation of blood circulation. Using a flexible tube, he connected the femoral artery of a horse to a long upright brass pipe with its top end remaining open. As soon as the clamp on the connecting tube was removed, the blood rushed from the artery into the brass pipe and started filling it until it rose to a level of about 2 m. The pressure of the blood column in the brass pipe was balanced by the arterial pressure (approximately 20 kPa). The blood level in the brass pipe was not constant — it oscillated at the frequency of heart contractions between the maximal (systolic) and minimal (diastolic) values. Systolic pressure corresponded to the contraction of heart, and diastolic pressure corresponded to its relaxed state.

He described this in the second volume of his works *Statical Essays: Containing Haemastaticks* (1733). Hales, however, was interested not only in the motion of liquids, but also in that of the air. And here, too, his ideas were translated into practice. To fight stuffiness in closed rooms, he was the first to suggest installation of windmill-like fans.

The technique for measuring arterial pressure, proposed by Hales, was associated with considerable loss of blood and even the risk for a patient. For this reason, the technique can be used to measure arterial pressure, perhaps, in experiments on animals only. An aspiration to develop a technique for measuring arterial pressure that would be fit for a human caused the Italian physician Scipione Riva-Rocci (1863–1937), in 1896, to invent the instrument still in use now (Figure 2.7). Typically, the instrument is employed to measure blood pressure in an artery of an arm. Since an artery in a lowered arm is at the level of the heart, the blood pressure in the artery coincides with that in the part of aorta, nearest to the heart.

The Riva-Rocci technique is based on measurement of external pressure required for pinching an artery. To this end, a hollow rubber cuff is placed on a patient's arm and, with the help of a pump, the pressure in it is raised

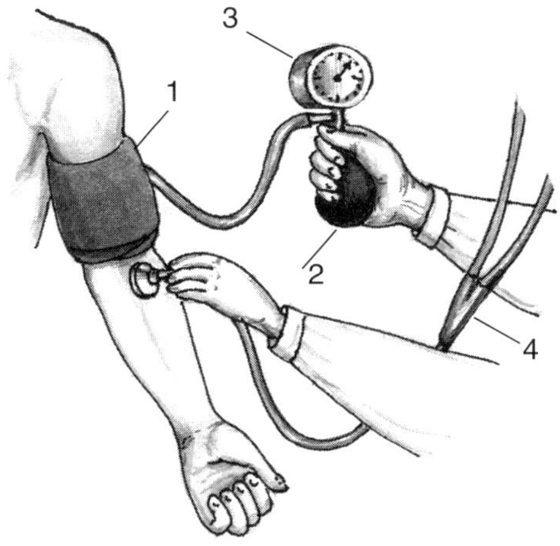

FIGURE 2.7. Riva-Rocci-Korotkoff technique for measuring arterial pressure in men: (1) cuff; (2) rubber bulb with a valve for pumping air; (3) gauge for measuring air pressure; (4) phonendoscope for listening to Korotkoff's tones.

until the pulse in an artery of a forearm (brachial artery) disappears. The air pressure in the cuff at the moment when the pulse waves in the forearm artery disappear (when the blood flow in it ceases) should equal systolic blood pressure.

In 1905, Russian physician N. S. Korotkoff modified the Riva-Rocci technique so that diastolic blood pressure could be measured as well. He suggested listening to the pulse waves in a forearm using a phonendoscope (the instrument comprises a sensitive membrane and two flexible tubes to take the sound vibrations to the eardrums).

If the air pressure in the cuff is raised above the systolic one and then lowered slowly with the help of a special valve, distinctive sounds appear at the pressure equal to the systolic value. The origin of the sounds, known as Korotkoff's tones, is related to a complex character of pulse wave propagation along a partially pinched artery. When the pressure in the cuff becomes smaller than the diastolic one, the artery starts passing blood unimpeded and Korotkoff's tones disappear. Therefore, the pressure in the cuff, corresponding to disappearance of Korotkoff's tones, is recognized as the diastolic one.

To get a full picture of the functioning of the cardiovascular system, it often does not suffice to measure the pulse rate and the arterial pressure. The unhealthy condition of some organ may be related to the diminished blood flow through the artery supplying it with blood. In these cases, to make a correct diagnosis, it is necessary to measure the velocity of blood flow through the artery (i.e., the volume of blood flowing through it in a unit of time).

Among the first to study the velocity of blood motion along vessels was the French physician and physicist Jean L. M. Poiseuille (1799–1869). Interestingly, the law that is named after him and that relates the velocity of the motion of a liquid through a capillary to the radius, to the length of the capillary and to a pressure differential, was the generalization of experiments performed by Poiseuille on the blood vessels of animals.

However, using Poiseuille's law to measure blood flow in the arteries of man is practically impossible, since this presupposes knowing the inside diameter of an artery, values of blood pressure at two points of it, and blood viscosity. Obviously, acquiring these data makes the technique rather "bloody" and often merely unrealistic.

Presently the velocity of blood flow through vessels is normally determined with the help of the instruments of two types: electromagnetic flowmeters and devices using the indicator-dilution method.

The electromagnetic flowmeter is based on the principle that when a conductor moves at right angles to the lines of force of a magnetic field, an electrical potential is induced. In this application of Faraday's law, the conductor is a stream of blood passing between the poles of a magnet. The induced current is led off by electrodes placed across the conduit of the stream. The current is proportionate to the velocity of the stream and its polarity is determined by the direction of the stream.

It is noteworthy that one of the creators of the theory of electromagnetic induction, Faraday, seeking to verify the validity of the law of electromagnetic induction for conducting liquids, wanted to measure the difference of potentials between opposite banks of the river Thames, which arises as its waters flow in the earth's magnetic field.

The application of this principle to blood flow was independently developed by A. Kolin in 1936 and E. Wetterer in 1937. Electromagnetic flowmeters represent a thin catheter with an outside diameter of 1–2 mm. This device can be introduced into many human arteries, with practically no change in the velocity of a blood flow in them. As a rule, the value of magnetic flux density inside the flowmeters is 10^{-3} T, and so the recorded electromotive force at normal blood flow speeds rarely exceeds 10^{-5} V. Nonetheless, despite such a small magnitude of the signal at the pick-up output, the electromagnetic technique has found wide use in clinical and laboratory studies.

The indicator-dilution method for the measurement of blood flow and volume arose from an almost two-century-old technique proposed by E. Hering in 1827, where potassium ferrocyanide was injected intravenously and the blood was collected at timed intervals from the corresponding contralateral vein. Hering tested for ferrocyanide by adding ferric chloride to serum. The first sample giving the Prussian blue reaction was, therefore, blood which had made one complete circuit from vein to heart, to pulmonary bed, to heart, to artery, to vein. The time at which this sample of blood was obtained (with a starting point set at the moment of intravenous injection) was called the *circulation time*.

The indicator-dilution method, further developed by G. N. Stewart in 1921, made it possible to determine not only circulation time but also the bulk velocity of blood flow, provided one knows the amount of indicator (dye, or other agent) injected into the blood and its concentration at a certain point. In most cases, indicators are various dyes that are harmless to organisms and noticeably different in color from blood. In such cases the indicator concentration in blood is established by photometry, whereby the tint of a vessel is measured in transmission. Sometimes the cooled physiological solution is used as an indicator. The concentration of such an indicator can be estimated by measuring the temperature of blood in the vessel.

The principle by which the injection of the indicator leads to a measure of flow is simple (Zierler, 1958). Over the years, two forms of injections have been used: in a *sudden-injection*, the indicator is injected very rapidly into the blood stream; in a constant-injection, the indicator is injected continuously at a constant rate.

Consider first the case of the constant-injection (Figure 2.8). Inject the indicator at a constant rate, I, into an organ with a fixed but unknown volume, V, available for blood perfusion. Blood flows through an organ at a constant but unknown rate F. After an equilibration period (see the shaded area in Figure 2.8(a)), the rate at which the indicator leaves the system will exactly equal the rate at which it is introduced into the organ. The rate at

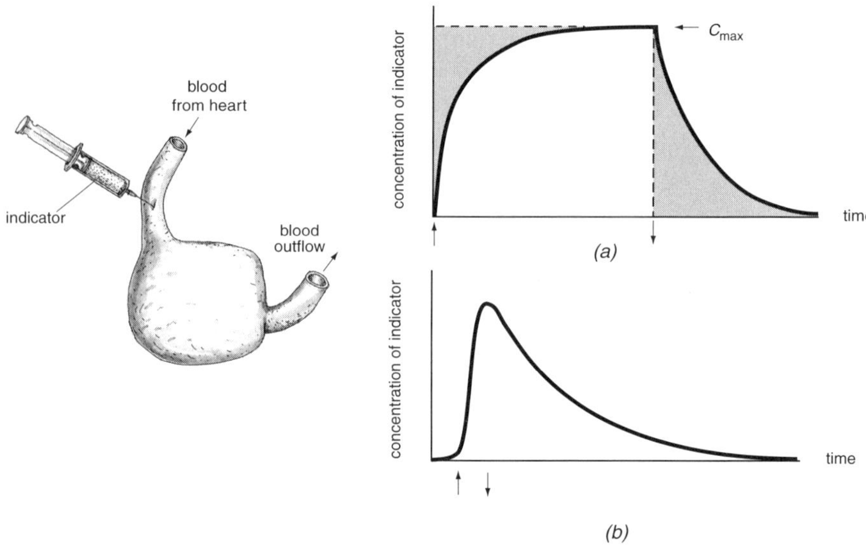

FIGURE 2.8. Principle of measurement of blood flow by indicator-dilution method. *Left:* schematic presentation of organ (or body part) through which blood flows at constant rate; indicator is injected into artery, which supplies the organ with blood. *Right:* concentration of indicator at organ outflow as a function of time during constant-injection, (a),or sudden-injection, (b). Arrows indicate the beginning and end of the indicator injection.

which it leaves the organ is a product of the measurable concentration of the indicator at the outflow from the organ and the unknown flow of blood, or

$$I = C_{\max} \times F \tag{2.13}$$

where $C_{\max}$ is the concentration of indicator at outflow. The relationship (2.13) makes it possible to compute F, provided I and $C_{\max}$ are known. It should be noted that the indicator-dilution method in described modification yields true values for the blood flow velocity in that case only when the indicator, after leaving a given section of a vessel or organ along with blood, is later withdrawn from the blood (e.g., by kidneys). Otherwise, the concentration of the indicator in blood would gradually grow and the computation of F according to Eq. (2.13) would result in too high values.

In the case of sudden-injection, a known amount of indicator is introduced into a blood vessel during a short time (about 1 second). In this case, the concentration c(t) of the indicator at the outflow from the organ will fail to be constant, but will vary in the manner similar to that shown in Figure 2.8(b). If velocity F of blood flow through an organ is assumed as constant, the amount of indicator leaving the organ during the time interval dt will equal $c(t)Fdt$. The total amount of indicator that has flown through the organ will equal $F\int c\ (t)dt$. If it is known and equals q, the equation $q = F\int c\ (t)dt$ yields

$$F = q/\int c\ (t)dt \qquad (2.14)$$

To find the volume, V, of an organ available for blood perfusion, first consider particles of blood entering the organ. The particles entering the organ at the same time require varying amounts of time to reach its outflow, the time required for any particle depending on the path taken and the velocity with which the particle travels. The blood (and indicator), therefore, does not have a single traversal time from inflow to outflow from the organ, but rather a distribution of traversal times. However, it is easy to come to the conclusion that

$$V = F \times t_{\mathrm{m}} \qquad (2.15)$$

where t_{m} is the mean traversal (or circulation) time which equaled $(F/q) \cdot \int tc(t)dt$.

The earliest technique for studying the blood flow velocity, still in use at present, is the one suggested by the German physiologist Adolph Fick in 1870 when he was just 26 years old. To establish the amount of blood, F, ejected by the heart in a unit of time (i.e., the velocity blood flow throughout the organism), he measured the concentration of oxygen in arterial (A) and venous (V) blood and also the amount of oxygen (O) consumed by an organism in a unit of time.

The amount of oxygen received by an organism from a unit volume of arterial blood is A–V. If the organism passes F blood volumes in a unit of time, the amount of oxygen consumed by organism is $F(A - V)$. On the other hand, this quantity (O) can be estimated by measuring the concentration of oxygen in an inhaled and an exhaled air. Since $O = F(A - V)$, we have

$$F = O/(A - V) \tag{2.16}$$

Again it should be noted, though, that the Fick technique is applicable to investigation of the velocity of blood flow through the heart only (total blood flow).

Widely used nowadays is the ultrasonic technique for measuring linear velocity of blood motion. The technique employs the well-known Doppler effect according to which the frequency of perceived sound vibrations is dependent on the velocity of the sound source motion relative to the sound receiver. Short pulses of ultrasound waves are transmitted diagonally across the blood vessel flow profile from one crystal and received by the other. Immediately, a second pulse originates in the second crystal and is received by the first. The average velocity of flow is derived from the time difference of sound transmission up and downstream. The frequency of the applied ultrasound is usually in the range between 1 and 10 MHz. The blood particles that scatter ultrasound and, therefore, act as its secondary moving sources are erythrocytes with sizes of about 5 μm.

Besides, some ultrasonic flowmeters measure the difference between the frequencies of transmitted ultrasound and that scattered by blood, which also makes it possible to compute the velocity of blood motion, given the position of a vessel relative to the pick-up and the speed of ultrasound in the medium. Despite an apparent ease of measuring the velocity of blood motion with the help of Doppler's principle, its application calls for use of special electronic equipment capable of detecting frequency changes about 0.001% of the transmitted one.

Note that the ultrasonic technique allows for finding just the linear velocity of blood motion, rather than the velocity of blood flow (see earlier). The latter, obviously, can be computed, if we multiply the velocity of blood motion by the cross-section area of a vessel. Unfortunately, in most cases the cross-section area of a blood vessel is difficult to evaluate with sufficient accuracy. In these cases the ultrasonic technique can provide us with the data only on the relative changes in the velocity of blood flow, with the cross-section area of a vessel assumed unchanged.

Blood Color and the Law of the Conservation of Energy

Well now, let us relax again before turning to the next chapter. The law of the conservation of energy in its clearest form was first stated in 1842 by the German physician and physicist Julius R. von Mayer (1814–78). This physical law was discovered under highly unusual circumstances. In 1840, Mayer went out to sea as a surgeon on board a Dutch ship to the faraway island of Java. At those times, the most widely used method of therapy was bloodletting, and it was common for a physician to see a patient's venous blood. And so, as they were approaching warm equatorial latitudes, Mayer noticed that the color of the venous blood of seamen became redder than it was in Europe. This signified that more oxygen is retained in human venous blood at southern latitudes than at northern ones.

It is obvious that the oxygen concentration in arterial blood is the same for different latitudes and depends only on its concentration in the atmosphere. This led Mayer to conclude that in cold climate men consumed more oxygen. He further deduced that maintaining the same body temperature in cold weather required a greater oxidation of food products.

However, Mayer realized that the energy released in combustion of food products was spent not only on maintaining constant body temperature in a man but also on his performing mechanical work. This meant that certain relationships should hold between the amount of heat generated in an organism and the mechanical work a man performs during a given interval of time. So Mayer concluded that a certain amount of heat should be in correspondence with a certain value of mechanical work delivered.

The idea of equivalence of heat and work attracted Mayer at once. Afterwards, Mayer's life went awry. There was a lot of argument around Mayer's priority in the discovery of the law of conservation of energy. Both this and troubles at home took their toll on the scientist's mind. In 1851, he was committed to a mental institution. Though after some time he was discharged, his contemporaries noted that Mayer's mind never recovered.

3

Crocodile Tears and Other Liquids

Amazing as it may seem, we are all more than half water. As Table 3.1 shows, the water content in every organ and tissue (except for fat and bones) lies in a range between 60 and 85%.

Normally, the ingress of water into an organism and its loss are balanced. In conditions of moderate climate, a person consumes 2.5 L of water a day on average; the corresponding water balance is described in Table 3.2. Thus, the daily water cycle of an adult is on the average about 3–4% of body mass. The minimal daily water demand of an adult is considered to be 1.5 L, 0.6 L of it necessary for removal of waste products by kidneys, with the remaining 0.9 L evaporating from the skin surface.

The greater part of substances removed by kidneys is comprised by urea (30 g, or about 0.5 M a day) and sodium chloride (10 g, or about 0.2 M a day). Thereby, the maximal concentration of osmotically active substances in urine amounts to 1200–1400 mOsm/kg H_2O, which determines the minimal volume of liquid discharged with urine.

The reduction of ingress of water into an organism (dehydration) is fraught with serious consequences. Some of them are related to the fact that

Table 3.1. Water content values for various organs and tissues.[1] (Reproduced by copyright permission of Academic Press.)

Tissue	Water Content, %
Skin	60–76
Muscle	73–78
Bone	44–55
Brain	68–85
Liver	73–77
Lung	80–83
Kidney	78–79
Spleen	76–81
Bone Marrow	8–16
Fat	5–15

[1]Data from Pethig (1991).

the blood in this case becomes more concentrated and viscous, consequently losing its ability to flow along the finest capillaries of most organs, so that such organs begin to die off. In cases when the amount of water in an organism diminishes by a third, death ensues.

It is noteworthy that cases of death caused by dehydration were observed in shipwrecked people who tried to quench their thirst with seawater, the osmolarity of which is about 900 mOsm/kg H_2O. The cause of dehydration is the fact that, in order to remove salts contained in seawater, the organism

Table 3.2. Daily water balance of adult human[1]. (Reproduced by copyright permission of Mosby Publisher.)

Water intake	mL/day	Water losses	mL/day
Drinking	1200	Urine	1400
Food	900	Lungs and skin	900
Metabolic processes[2]	300	Feces	100
Total	2400	Total	2400

[1]Data from Muntwyler (1968).

[2]Oxidation of 1 g of hydrocarbons, fats, and proteins yields 0.6, 1.1 and 0.4 ml of water, respectively.

is forced to use its own water, with the consumption of 1 L of seawater accompanied by the formation of at least 1.6 L of urine.

But then, how do sea birds quench their thirst? In 1939, the well-known Norwegian physiologist Knut Schmidt-Nielsen decided to find the answer to this question. He caught several arctic gulls, placed them in cages, and started feeding them with fresh sea fishes. Under such a diet, the gulls showed no signs of thirst and Schmidt-Nielsen inferred that the birds received all required water with the caught fish, in which the concentration of salts was much lower than in seawater.

Well then, what if the fish swam to the sea bottom as is the case during a sea storm? Then, obviously, the gulls have no choice but to drink seawater. To learn how gulls discharge salts out of an organism in that case, Schmidt-Nielsen started to force them to drink seawater, expecting to observe an abrupt increase in the concentration of salts in their urine. However, it turned out that such a salt diet, by all means, failed to produce an increase in the concentration of salts in the urine of gulls. Instead, they take to "weeping." Faced with the excess of salt, special glands near the beak squeeze out a liquid in which the concentration of salts can be several times higher than that in seawater. Apparently, the same happens with crocodiles, who often inhabit salty water, so that the "crocodile tears," which became a synonym of hypocritical compassion, could be another means of discharging excessive salts out of an organism. However, for obvious reasons, nobody ever studied crocodile tears in detail. Human tears also contain salts; however we are, of course, unable to "weep out" all the salts that enter our organism daily. That is why we can rely only upon the functioning of our kidneys.

Water in Us

How can we find out how much water there is inside us? Usually the techniques based on the principle of indicator dilution are used for this purpose. It is evident that if a person's blood is injected with a harmless substance, which will freely (the same way as water) permeate the membranes of all cells of an organism, then after some time its concentration will become the same throughout all liquid phases of the organism. After that, the volume of the liquid phase of the human body can be determined by dividing the amount of injected indicator to its concentration in the organism.

To establish the total volume of water in an organism, antipyrine, as well as heavy water (D_2O or ^3H_2O), is most often used as an indicator. In two hours' time, these substances become uniformly distributed among various liquids of the organism. During that time, however, a fraction of the injected substance is withdrawn from the blood bed and is concentrated in the bladder, which interferes with the evaluation of a true volume of liquid phase of an organism. Therefore, a mathematical model is needed to provide a qualitative approximation of the indicator dilution process.

Let V be the volume of a liquid phase of organism; $C(t)$ be the concentration of indicator in it; ΔC be the change of the concentration during the time period Δt and v_O be the volume rate of discharge of liquid together with the dissolved indicator through the kidneys. Then the law of conservation of mass (for indicator) yields

$$V\Delta C = -Cv_O \cdot \Delta t \tag{3.1}$$

By integrating Eq. (3.1), we get

$$\ln C(t) = \ln C(0) - tv_O/V \tag{3.2}$$

where C(0) is the concentration of indicator immediately after its injection into an organism (assuming its penetration into all liquid media and its mixing occur instantaneously).

It follows from Eq. (3.2) that the curve $C(t)$ in semilog coordinates should have the form of a straight line intersecting the axis of ordinates at point C(0). Thus, if the experimental indicator dilution curve (i. e., the dependence of indicator concentration on time) is plotted in semilog coordinates and fit to the axis of ordinates (Figure 3.1), we can obtain the desired value C(0) required for computation of the volume of the liquid phase of an organism.

The actual measured indicator concentration (solid line) in the left-hand part of the plot in Figure 3.1 is noticeably higher than its fit (dashed line). This is evidence that in reality the uniform distribution of the indicator fails to take place instantaneously, and that during all this time its concentration in blood somewhat exceeds the one in other liquid media of the organism.

In adults, the mass of water contained in the organism and measured in

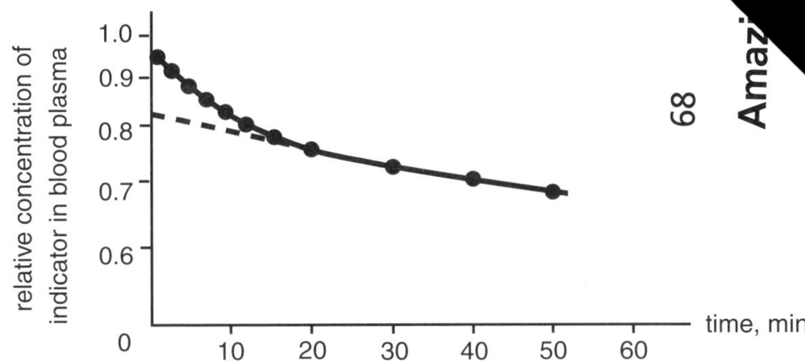

FIGURE 3.1. Determination of the volume of liquid with the help of indicator dilution technique. Plotted is the variation of indicator concentration (*C*) in patient's blood plasma upon a single (as a single bolus) injection of indicator. (Modified from Pitts, 1974.)

the described manner on the average amounts to 60% of body mass for men and 50% for women, with a greater part of water being found in muscles (32% of body mass), skin (13%), and blood (7%).

A similar method can be used to measure the volume of extracellular liquid in an organism: a person's blood is injected with indicators incapable of penetrating through cell membranes. Such indicators are typically various sugars, while inulin, which is removed from an organism very quickly, was selected by researchers as a standard substance to be used for determining the total extracellular space (see Figure 3.1).

The mass of water in extracellular space found with the help of inulin (inulin space) on the average amounts to 16.5% of body mass. However, if we use a substance that is removed from the organism slowly (e.g., thiocyanate) as an indicator, and wait for a long enough time (from 5 to 10 hours), it will turn out that the true extracellular space may amount to 27% of body mass.

The volume of water contained inside cells can obviously be found by subtracting the extracellular fraction from the total water contained in an organism. Therefore, the intracellular water is believed to amount to about 33% of a person's mass.

ing Filter

l the cells of our organism are surrounded by intercellular liquid on every side. A necessary condition of a cell functioning is the constancy of the volume and composition of the liquid. This statement was first advanced more than 100 years ago by the well-known French physiologist Claude Bernard (1813–1878). In what way is this constancy maintained?

In principle, the amount of water in our organism and the composition of intercellular liquid are regulated by us subconsciously when, for instance, we appease the feeling of hunger or thirst, since we receive water and electrolytes mostly with eating and drinking. An exception is the so-called metabolic water forming in the oxidation of food products (see Table 3.2). The greater part of the excessive water and electrolytes is removed through the kidneys. Therefore, the base of maintaining the constancy of the volume and composition of liquids in an organism is the normal functioning of kidneys.

A functional structure unit of a kidney, which is the site of urine formation, is the *nephron* (Figure 3.2). Each human kidney weighing about 150 g contains approximately one million nephrons.

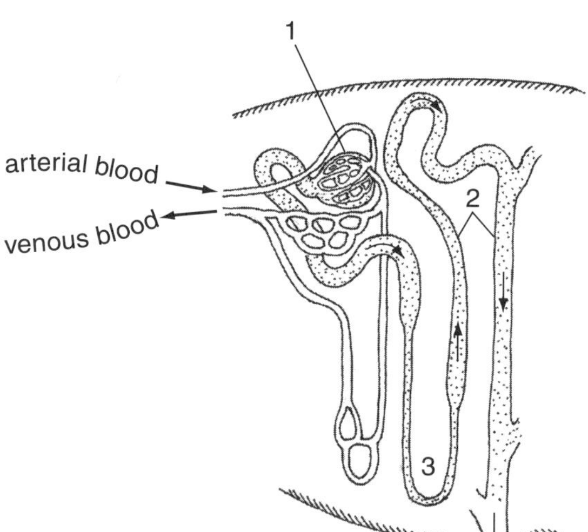

FIGURE 3.2. Schematic presentation of a nephron and its blood supply: 1, renal corpuscle; 2, renal tubule bent in the form of 3, Henle's loop.

A nephron is made up of two main parts: renal corpuscle and tubule (descending and ascending limb, marked 1 and 2, respectively, in Figure 3.2.) Blood, passing along capillaries situated in the corpuscle, filters through their walls and the resulting filtrate finds itself in a cavity opening into a tubule. This liquid, which is already free from molecules with molecular weight in excess of 80 kD, became known as primary urine. Its daily volume amounts to about 180 L; it differs in composition from blood plasma solely by the absence of high-molecular proteins.

The source of energy for blood plasma filtration in a renal corpuscle is the work of the heart. The heart, when contracting, produces excess pressure (20–30 mm Hg) inside a corpuscle capillary, which is what forces some part of blood passing along the capillary to filtrate through its wall and form primary urine.

In a renal tubule bent in the form of Henle's loop (marked 3 in Figure 3.2), concentration of primary urine takes place. As a result, 99% of water (178.5 L a day) returns back into the blood and mere 1.5 L is discharged in the form of urine. Here, the osmolarity of urine may be as high as four times that of blood plasma, which is typically in the range between 285 and 295 mOsm/kg H_2O. Let's consider in greater detail the way the concentration of solution in Henle's loop occurs.

The first model to explain the mechanism of the solution concentration in Henle's loop was proposed by Kuhn and Ryffel (1942) and elaborated later in the work of Hargitay and Kuhn (1951). Kuhn and his coworkers suggested that the Henle's loop provided a countercurrent system in which a concentration of solutes could be multiplied manifold. To model Henle's loop, they presented it as a tube divided with a semipermeable membrane *(M)* into two limbs (left, *L*, and right, *R*) of the same size (Figure 3.3). The loop limbs are interconnected via a narrow capillary tube (*C*), along which the liquid flows from the left limb into the right one under the action of pressure.

First, let the capillary connecting the loop limbs be closed and both limbs be filled with liquid of the same composition (Figure 3.3(a)). Naturally, in this case there will be no movement of liquid along the loop. However, if hydrostatic pressure is applied to limb *L*, the water (for which the membrane is permeable) starts moving from the left limb into the right one. As a result, the concentration of substances in the left limb will grow, and in the right, decrease.

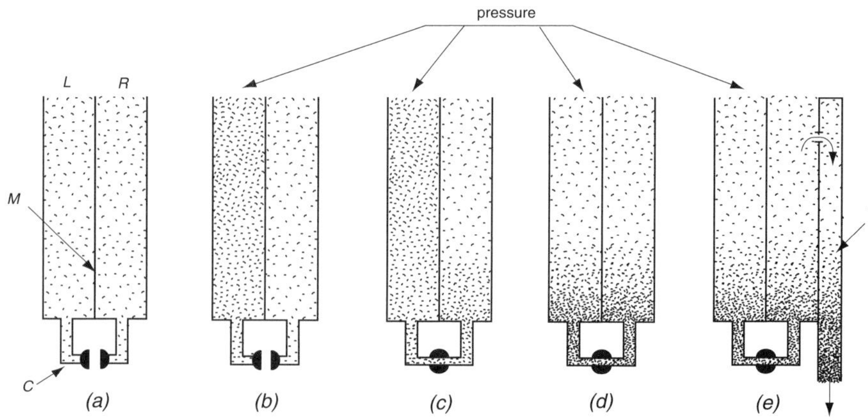

FIGURE 3.3. Illustration of countercurrent mechanism of urine concentration in Henle's loop.

Still, as soon as the concentration of substances in the left limb of the loop starts growing, a reverse flow of water (from right to left) arises, the flow caused by the osmotic pressure gradient.[1]

It can be shown that the gradient of osmotic pressure $\Delta\pi$ is computed according to formula

$$\Delta\pi = RT(C_{\mathrm{L}} - C_{\mathrm{R}}) \tag{3.3}$$

where R is the gas constant, T is the absolute temperature, and C_{L} and C_{R} are the concentrations of solutes in left and right limbs of the loop, respectively.

It is obvious that when the left-to-right water flow, caused by the hydrostatic pressure gradient, becomes equal to the osmotic flow from right to left, an equilibrium will be attained (see Figure 3.3(b)). It will occur when the osmotic pressure gradient becomes equal to the hydrostatic pressure

[1] Osmos is the term used to refer to the transport of solvent (in this case, water) through a semipermeable membrane separating two solutions with different concentrations. Here, the solvent molecules pass through the membrane, impermeable for solutes, into a more concentrated solution. The process runs until the concentrations are balanced.

applied to the left limb of the loop; the corresponding difference of concentrations can be found with the use of Eq. (3.3).

Now, let us open capillary *(C)* connecting the loop limbs. Since the capillary is very narrow, the hydrostatic pressure gradient between the limbs can be considered to remain unchanged. At the same time, immediately after opening the capillary (at first, in the lower part of *R* and later in the upper one, too) concentrated liquid will appear (see Figure 3.3(c)). This means that the equilibrium between *L* and *R* is disturbed (the gradient osmotic pressure has decreased) and the water ingress from left to right will recommence. As a result, the concentration of substances in the liquid of the left limb near the capillary is growing. Thus, the countercurrent system of liquids exchanged through a semipermeable membrane leads to the concentration of solution near the turning point (Figure 3.3(d)).

It is obvious that in the state of equilibrium the concentration of a solution that outflows from the loop (see Figure 3.3(d)) is the same as that of the one that enters. Consequently, the capability of a countercurrent loop for concentration is not used in this case. To remove the concentrated solution from Henle's loop, nature provided for the *third* (*T*) limb. In accordance with the model of Kuhn and his coworkers, the limb is separated from *R* by a semipermeable membrane and connected to *R* by a small opening (Figure 3.3(e)). Since the opening is very small, only a small fraction (about 1%) of liquid flows out of *R* into limb *T*, and for this reason, the motion of the liquid along the loop and the concentration gradient are not disturbed. At the same time, the liquid, which moves slowly top-down along limb *T* (the diameter of the limb is the same and the liquid flow rate is about 1%), attains osmotic equilibrium through the semipermeable membrane and exits the three-limb Henle's loop with a very high concentration of substances, which equals the one at the junction of *R* and *L*.

Let us try to estimate the capability of Henle's loop for concentration. Divide each loop limb vertically into *N* segments: the first one at the top and the *N*th one at the very bottom (Figure 3.4). Let the concentration in the left loop limb in *k*th segment at moment *t* be equal to $L_k(t)$, and in a similar segment on the right, equal to $R_k(t)$.

To simplify the computations, let us assume that the solution motion along the loop has an intermittent character. Thereby, the solution instantly moves over by the length of segment *s*, staying at rest afterwards for time interval s/v where v is the average linear velocity of the liquid motion along

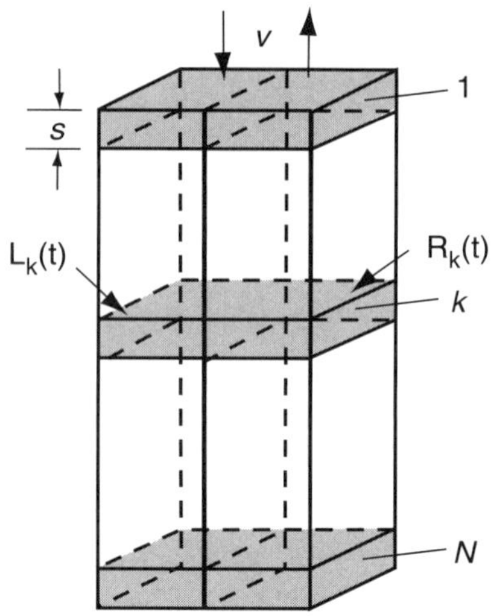

FIGURE 3.4. Elements of Henle's loop model.

the loop. We shall consider the transport of water between the neighboring limb segments through a semipermeable membrane (fully permeable solely for water) to occur only upon completion of the next-in-turn step and to last for the time period equal to s/v.

The process of solvent transport (in this case, water) through a membrane under the action of hydrostatic pressure gradient is referred to as *ultrafiltration*. Volume rate $\Delta V/\Delta t$ of ultrafiltration can be determined from the equation

$$\Delta V/\Delta t = k_f A(p + \Delta\pi) \tag{3.4}$$

where k_f is the filtration factor, A is the membrane area, p is the magnitude of hydrostatic pressure gradient, and $\Delta\pi$ is the osmotic pressure between these solutions, the pressure related to the concentrations of substances in them by Eq. (3.3).

Let all loop segments be identical and have the shape of a cube with lateral face area A. Then, substituting $\Delta t = s/v$ and assuming $V = sA$, we obtain

$$\Delta V/V = k_f\{p + RT(R_k - L_k)\}/v \tag{3.5}$$

Expression (3.5) enables us to compute the relative change of the water amount after a single ultrafiltration interval with duration s/v. Evidently, knowing this we are already able to evaluate the changes of concentrations in both adjoining segments after a single act of ultrafiltration:

$$\begin{aligned} L_k \text{ (after) } &= (1 + \Delta V/V) L_k \text{ (before)} \\ R_k \text{ (after) } &= (1 - \Delta V/V) R_k \text{ (before)} \end{aligned} \tag{3.6}$$

Well, we seem to have described the ultrafiltration in full, but the liquid also moves. Let us write down kinematic relationships. Let our unit of time *(t)* be s/v. In our model, the solution that has just entered the left kth segment is of the same osmotic concentration as the (k–1)th segment at the previous moment in time (similar dependencies are valid for the right limb segments). Therefore, kinematic relationships will have the form

$$\begin{aligned} L_k(t+1) &= L_{k-1}(t) \\ R_k(t+1) &= R_{k+1}(t) \end{aligned} \tag{3.7}$$

Primary urine with a constant osmotic concentration, a, keeps entering the first segment of the left part of the loop; hence, in computations using Eqs. (3.6) and (3.7) we should let $L_1 = L_1(\text{before}) = a$. In addition, for computations using Eq. (3.7) we should, obviously, let $L_0 = a$.

The liquid enters the lowest segment of the right limb of the loop immediately from the lower segment of the left limb, bypassing the capillary, its volume being negligible. Therefore, R_{N+1} in Eq. (3.7) should be assumed equal to L_N.

Simultaneous equations (3.5)–(3.7) describe the change of osmotic concentrations in transition from t to $t + 1$. To solve the equations, it is necessary to set initial conditions; that is, the values of variables at $t = 0$. Assume that at instant $t = 0$ the loop is filled with primary urine having osmotic concentration a, but there is no motion of the solution (hydrostatic pressure is not applied to the left limb). Then, evidently, we should let

$$L_k(0) = R_k(0) = a \tag{3.8}$$

The set (3.5)–(3.7) with initial conditions (3.8) is solved rather easily with the use of PC. If we let $N = 50$, $a = 300\,\text{mOsm/kg H}_2\text{O}$, $p/RT = 100\,\text{mOsm/kg H}_2\text{O}$, $k_f = 2 \cdot 10^{-10}\,\text{Pa}^{-1}\text{s}^{-1}$, $v = 0.00004\,\text{m/s}$, $RT = 2.24\,\text{kPa/mOsm/kg H}_2\text{O}$, it turns out that as time goes on, L_N gradually attains a steady-state value that equals (in Osm/kg H_2O):

t	10	20	50	100	200	500	1000	2000	5000
L_N	0.38	0.43	0.53	0.65	0.83	1.16	1.38	1.44	1.45

Of course, it is impossible to determine exactly the factors involved in this set of equations. Therefore, the computation results should be taken solely as an illustration of the physical process running in the countercurrent loop, with attention paid to basic patterns only.

Figure 3.5 shows the values the solution concentration *(C)* takes on in different segments *(k)* of Henle's loop under such simulation. It follows from the data obtained in simulation that the countercurrent mechanism of concentration can increase the solution concentration at the turning point of the loop fivefold. Note that ultrafiltration by itself (without countercurrent)

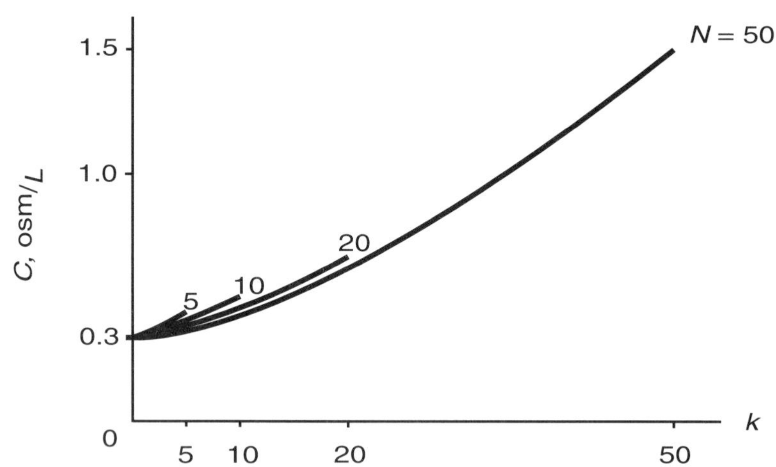

FIGURE 3.5. Dependence of solution osmolarity (*C*) on sequential number (*k*) of segment, evaluated for model in Figure 3.4. Plotted are the curves for Henle's loops of different lengths (N = 5, 10, 20, and 50).

is able to increase the solution concentration by mere 50 mOsm/kg H_2O (by 17%).

Our model allows for assessment of the dependence of the capability of the loop for concentration on its length (see curves in Figure 3.5 for $N = 5$, 10, 20, and 50). As we should expect, the longer the loop, the greater its capability for concentration.

It is obvious that for mammals living near freshwater reservoirs, and sometimes right in them, water is no problem. There is small need for them to save water. For this reason, the amount of water in urine may be large and, hence, the concentration of substances in urine (osmolarity), low. On the contrary, for mammals living in deserts, far from water bodies, every drop of water counts. Therefore, removing waste products with urine, desert inhabitants must raise its osmolarity as high as possible.

Indeed, Schmidt-Nielsen and O'Dell (1961) demonstrated that the osmolarity of urine in beavers amounts to less than 0.5 Osm/kg H_2O, whereas in the desert rodent *Psammomys* it exceeds 4 Osm/kg H_2O. The difference in osmolarity of urine in these animals can be accounted for by the fact that Henle's loop in the desert rodent *Psammomys* is almost 10 times longer than that in beavers.

Of course, during the past 40 years the model proposed by Kuhn and coworkers has many times come under criticism for oversimplification. Therefore it was followed by more involved models, which take into account recent experimental data (Stephenson, 1973; Layton, 1986; Weinstein, 1994). However, like in the model of Kuhn and coworkers, in all the new models, the crucial part still belongs to Henle's loops, which act as a countercurrent multiplier.

Cryobiology and Biological Antifreezes

Life is possible in a very narrow temperature range, from several degrees below freezing (0°C) to 40–50°C. Of course, it is the body temperature, not the temperature of the environment that matters here. Temperature fluctuations radically affect many physiological processes. Temperature reduction slows metabolism to such an extent that a 10 degree temperature decrease entails a 2–3 times decrease in the rate of physiological processes. The latter observation is used widely in a long-time storage of certain isolated organs and fluids of a man. For example, sperm tanks are stored in liquid nitrogen at –196 degrees C. And every year thousands of cancer patients receive transplants of their own frozen bone marrow, to replace the cells destroyed by radiation or chemotherapy.

Having found themselves outside the range of temperatures compatible with active life, many animals are capable of surviving by passing into a state of torpor or hibernation. However, such resistance with respect to low temperatures is not exhibited by all organisms. Anyone who keeps tropical fish in indoor fish bowls knows, perhaps, that once the heating is disconnected, the first cool night will kill all the fish.

It has been established that the cause of the death of a cell due to freezing is the formation of ice crystals *inside* a cell. The ice crystals destroy intracellular structures, and, as a consequence, the cell dies. Therefore, to save a cell from "crystal" death in low temperatures, the cell must be freed from the source of crystals—water—replaced (at least partially) with another liquid.

Glycerin has long been known to protect living organisms against damage in freezing. Hemolymph of insects contains glycerin in a high concentration (several tens of mM/l) and their survivability at low temperatures is attributed to this. Thus by the onset of winter, glycerin concentration in wasps increases several times, and in this period glycerin is responsible for about 3% of the total liquid content of the insect. As a result, the freezing point of hemolymph of wasps falls down to –17.5 degrees C. The same explanation apparently holds for the recent entomological discovery. In one of the Himalyan glaciers they found a mosquito-like insect who was fairly active at temperatures below –16 degrees C.

It has been proven that the ability to survive abrupt onset of cold by way of increasing the glycerin content in blood is typical not only of hibernating insects. Thus, common meat flies easily survive temperatures as low as –10 degrees C. However, they survive only under a relatively gradual onset of cold. Thus, whereas abrupt (several seconds long) freezing to –10 degrees C is lethal for all insects, practically all of them can survive gradual freezing (0°C for two hours, and then –10°C).

It turned out that the reason for such a fast adaptation of insects to the freezing temperatures is a threefold increase in glycerin concentration in their hemolymph. This, apparently, helps them survive during frosts in early spring and late fall.

The feature of glycerin as a good cryoprotector finds wide use in biology and medicine. It is known that erythrocytes can be stored in the frozen state for many months without damage, provided they are immersed in glycerin. A similar method can be used to protect even entire animals against cryodamages. Thus in 1992, research workers from the cryogenic company BioTime in Berkeley, California, cool a baboon down to about 2 degrees C and preserve him in this state for 55 minutes using a patented cold-resistant substitute, similar to glycerin in its properties. After this, the monkey was refrozen without any apparent consequences. You can learn of other semi-fantastic experiments on freezing animals from the paper by J. Knight (1998).

In contrast to mammals, ectothermic vertebrates (fish, snakes, frogs, turtles, lizards) that live in conditions of cold climate endure long cold winters, when their temperature may fall far below the freezing point of water. Nature employs at least two methods to protect these animals against death under such conditions:

1. Fish of subpolar seas synthesize antifreeze glycopeptides or antifreeze peptides (AFP) preventing ice crystal growth down to – 2.2 degrees C or below, which is substantially lower than the temperature of ice-laden seawater (DeVries,1988).
2. Other ectothermic vertebrates inhabiting the land have developed the ability to endure the freezing of extracellular fluids to such an extent that during wintering as much as 65% of their total body water is locked in ice (Storey and Storey, 1992).

The fish swimming in cold subpolar waters (winter flounder, Alaskan plaice, and shorthorn sculpin) exhibit a unique capability not to freeze, remaining in the supercooled state down to temperature –2.2 degrees C. In comparison, most fish of tropical and midlatitudes freeze, with ice present, when cooled down to –0.8 degrees C. The comparison of the two figures would make some smile — a mere 1.4 degrees?! Yes, it is these one and a half degrees that help Antarctic fish survive. Indeed, at inlet McMurdo Sound (the part of the ocean nearest to the South Pole), for instance, the average annual temperature is –1.87 degrees C with variations from –1.4 to –2.15 degrees C. However, the mechanism used by winter flounder to avoid freezing when swimming among ice is different from that used by insects.

Before we learn the secret of subpolar fish, let us see upon what the formation of ice crystals depends. It has been established that freezing points of most solutions depend on the amount of dissolved particles, rather than their nature. The presence of dissolved particles obviously diminishes the probability of formation of a nucleating center, since the number of collisions of water molecules with one another goes down. For instance, such is the action mechanism of salt solutions, still in use in many cities to fight icing. Glycerin, which precludes freezing of insects in a cold season, apparently acts in a similar way.

Yet another, more refined action mechanism of antifreezes is possible, one that does not require their high (sometimes, molar) concentration. It turned out that some polypeptides and glycoproteins, whose molecules have the helical shape and consist of a multitude of repeating units, and whose molecular weight ranges between 3000 and 30,000 D, are capable of lowering the freezing point noticeably already in millimolar concentrations. And if these protein antifreezes are compared to NaCl, the former prove to be 300–500 times more efficient. What is the action mechanism of antifreeze peptides?

Molecules of water in ice crystals form hexagonal lattice with oxygen atoms at vertices of hexagons. Therefore, under the ideal conditions the ice crystals are hexagonal pyramids.

It was established (Knight *et al.*, 1991) that numerous polar groups in molecules of antifreeze peptides, extracted from winter flounder and Alaskan plaice and capable of forming hydrogen bonds with water molecules, are located at the same distance from each other (16.7Å) as the water molecules along one of their directions $\{01\bar{1}2\}$ in pyramidal planes of ice. As a consequence, long molecules of antifreeze peptides can considerably slow down the crystal growth by binding to the fast-growing end-face.

Antifreeze peptides are responsible for about 3.5% of mass of all liquids in a polar fish body. It is these antifreezes, acting jointly, that lower the freezing point by approximately 1.2 degrees C. Lowering it by still another degree is due to various ions and molecules (mostly NaCl) contained in liquids of Antarctic fish. The concentration of antifreeze peptides in liquids of warm-water fish is negligible.

The study of the binding mechanisms of antifreeze peptides to ice surfaces potentially has numerous practical applications. For example, it is crucial in cryopreservation research since ice crystal growth can significantly damage cryopreserved biological material. Therefore, adding the antifreeze peptides to a protectant solution could reduce damage due to ice crystal growth. Further practical applications include prevention of damage to agricultural crops by early frost (Kenward *et al.* 1993) and inhibition of ice crystal growth in foods stored at low temperatures (Feeney and Yeh, 1993).

Extracellular liquid in animals and plants freezes earlier than does intracellular liquid. Under slow freezing of extracellular salt solution, water crystallizes, whereas salts accumulate between crystals, raising the osmotic concentration of the remaining extracellular solution. As a result, concentrated extracellular solution sucks the water out of cells, and they are dehydrated. The freezing point of intracellular solution lowers. Thus, if conditions are provided for prevailing formation of ice crystals in extracellular liquid, the cells proper can be saved from freezing. Ectothermic vertebrates wintering on land resort to this way of salvation.

Experiments performed on wood frogs, *R. Sylvatica* (for review, see Storey and Storey, 1992), showed that the frogs can endure a decrease of body temperature down to –4 degrees C, lasting for several days, whereby the frogs consist of ice by more than a half of their body. Thereby, ice crystals appear only in extracellular liquids of frogs, rendering the freezing process reversible. Reptiles and amphibians have the ability to supercool (chill below the freezing point without freezing). Ice-nucleating proteins, appearing in blood immediately before wintering, initiate and control the formation of ice in extracellular fluids of freeze-tolerant animals. The action of frog ice-nucleating proteins results in a mean supercool point of –6 degrees C compared to –16 degrees C for human blood plasma. Amazingly, the addition of mere 0.5% vol/vol of cell-free frog blood raised the supercool point of human plasma to –7 degrees C, suggesting that frog ice-nucleating proteins might be used as effective agents in cryomedical systems. Note that in spring and summer reptiles and amphibians stop synthesizing the ice-nucleating proteins and lose this freeze-tolerant ability.

However, it is not always the case that the presence of ice-nucleating proteins that facilitate the ice crystallization helps organisms survive under lowering of the temperature; sometimes it becomes the very cause of their death. Such is, for instance, the role of some kinds of proteins discovered on the outer membrane of bacteria *Erwinia herbicola, Pseudomonas syringae*, and others. These bacteria, which can usually be found on surfaces of plants in Europe, Asia and North America, belong to a harmful category, and are considered to cause the low cryoresistance of plants (Arny *et al.* 1976) to them.

It is known that even plants, very sensitive to cold, can endure lowering of the temperature down to several degrees below 0 degrees C because of supercooling of intracellular water. Cryodamages of such plants outdoors occur at temperatures between –2 and –5 degrees C, and result from the ice crystal growth from supercooled intracellular water.

However, if the same plants are grown under sterile conditions, precluding ingress of bacteria onto their surfaces, then even cooled down to –8 degrees C, no crystallization of intracellular water (thus, no damage) occurs. The treatment with antibiotics (streptomycin or tetracycline), by killing bacteria, also substantially helps improve the frost resistance of plants.

Scientists have established that the protein contained in the membrane of these bacteria possesses a unique property to bind water molecules by assembling them into a configuration, similar to that present in ice crystals. As a result, microscopic crystals appear on the membrane of bacteria, and serve as nucleating centers for crystallization of all intracellular water.

The capacity to facilitate ice crystallization has been shown to be present but in very few species of bacteria. Thus, out of 42 species of bacteria collected from hawthorn leaves, just one species, *Pseudomonas syringae van Hall*, exhibited these properties. Yet, not every bacterium of the species contains the unique protein-water crystallizer in its membrane.

It is believed that the bacteria that serve as nucleating centers for ice crystallization can have a crucial role in setting the climate of an area by controlling the temperature of crystallization of atmospheric moisture.

4

Inhale Deeper

"Pure air, in passing through the lungs, undergoes then a decomposition analogous to that which takes place in the combustion of charcoal," Antoine L. Lavoisier (from Leicester and Klickstein, 1952). Indeed, what but a continuous combustion can account for the fact that our body temperature is constant and almost always higher than the ambient one? Lavoisier believed that the "stove" which warms a man is found in the lungs. He claimed that, like in an ordinary stove, the carbon of the living tissue undergoes a chemical reaction with the oxygen in the air to yield carbon dioxide, and as a result, the necessary heat is released.

In reality, the reaction involving oxygen occurs not only in the lung cells, but also in all the cells of an organism, where oxygen is delivered by blood. Besides, the process involving oxygen and supplying us with energy (including heat) has nothing in common with the reaction of the direct combustion of carbon, but rather is a long chain of chemical reactions yielding CO_2 as one of the end products.

Yet for simplicity, our organism can sometimes be regarded as a stove that uses about 0.5 kg of oxygen a day, and which gives off an almost identical amount of carbon dioxide. To further draw on Lavoisier's analogy, the lungs in the stove are assigned the part of an ash-pit for oxygen to enter and a pipe for carbon dioxide to escape.

Interestingly, in the seventeenth century the well-known English physicist Robert Boyle (1627–1691), who discovered one of the laws of gases, stated that in passing through the lungs the blood "is freed from harmful evaporations." How do lungs, which take up only about 5% of our entire body volume, manage to fulfill the task?

The inner space of a lung communicates with the atmosphere via ducts that comprise the nose, where inhaled air is warmed up and humidified; larynx; trachea; and two bronchial tubes, which feed air to the right and left lungs. Each bronchus may undergo 15 or more branchings, splitting into smaller bronchi, to end with microscopic follicles (alveoli) shrouded in a dense network of blood vessels.

Alveoli, of which there are about 300 million in an adult, represent bubbles filled with air. The average diameter of alveoli is roughly 0.1 mm and their walls are 0.4 μm thick. The total surface of alveoli in the lungs of man amounts to about 90 m^2.

At any single moment, the blood vessels enlacing alveoli hold approximately 70 mL of blood, where carbon dioxide diffuses into alveoli, and oxygen diffuses in the reverse direction. Such enormous surface of alveoli makes it possible to diminish the thickness of the blood layer that exchanges gases with intra-alveolar air down to 1 μm, which allows for saturating this amount of blood with oxygen and ridding it of the excess of carbon dioxide in less than 1 second.

It should be noted that the human in-breath is effected not only by lungs, but also by the entire body surface — the skin, from head to toes. Especially intense is the breathing of the skin in the breast, back, and abdomen areas. Interestingly, these sections of skin are much superior to lungs in the intensity of breathing. Thus, for instance, a unit of the surface of such skin can adsorb 28% more oxygen and give off 54% more carbon dioxide than lungs.

This superiority of skin over lungs may be due to the fact that our skin breathes clean air, whereas our lungs are ventilated poorly (see the following section "Exceptions to the Rule"). However, the share of skin in man's breathing is negligible as compared to lungs since the total surface of the human body falls short of 2 m^2, thus constituting less than 3% of the total surface of the lung alveoli.

When we inhale, the volume of air in our lungs increases as the air from the atmosphere comes in. Since alveoli are the most elastic part of a lung, practically all changes in lungs volume under inhaling and exhaling occur as a result of the respective changes in alveoli volume. In inhaling, alveoli expand and in exhaling they shrink. About 15,000 times a day we extend the alveoli of our lungs, whereby we deliver mechanical work that comprises between 2 and 25% of all our energy outlay. What determines the amount of this work?

The work we deliver in breathing is spent on overcoming the forces of resistance of several types. The first and the most sizeable part is spent on extending the lungs. The second one is the work spent on moving the air in the direction of alveoli. The air flow may be either laminar or turbulent, and the energy expenses depend on which of the two it is.

The structure of the nasal cavity is conducive to the turbulent flows of inhaled air. As is the case in a centrifuge, turbulent air flow yields a more efficient warming up of the air and separation of alien particles. Turbulence of inhaled air flow also arises at numerous sites of bronchial tree branchings.

Since laminar gas motion transforms into the turbulent one with the increase of the air flow velocity, obviously, the relative role of resistance forces in breathing will depend on the frequency of the latter. It has been shown that the frequency at which we normally breathe (about 15 breath intakes a minute) is consistent with the minimal energy outlay for breathing.

Breathing and Soap Bubbles

In 1929 the Swiss physician Kurt von Neergaard demonstrated that the pressure required to distend the lungs could be radically reduced if lungs were filled with the salt solution, which is close in composition to intercellular liquid (Figure 4.1).

If every alveolus is considered to be a hollow ball enclosed with an elastic membrane, the air pressure required to sustain the ball in a distended state should be fully determined by the ball diameter, membrane thickness, and Young's modulus, but should not depend on what the ball is filled with. Von Neergaard came to the conclusion that the alveoli normally have a wet lining and therefore the force of surface tension must add to their elastic recoil.

To assess the role of surface tension in mechanics of alveoli, consider a sphere-shaped film of liquid. As in the case of a flat film, the surface tension forces here seek to reduce the surface of the sphere by compressing the air

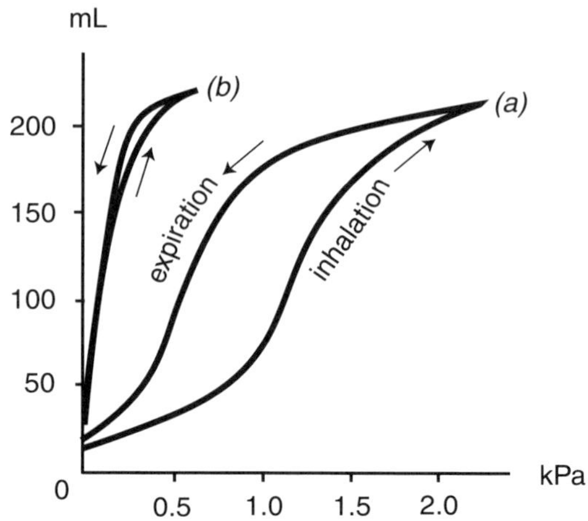

FIGURE 4.1. Pressure-volume curves for cat lungs filled with air (a) and an aqueous solution (b). (Modified from Radford 1954. Reproduced by copyright permission of Academic Press.)

inside it. As a result, the air pressure inside the sphere formed by liquid film always turns out to be somewhat greater than the atmospheric one. According to the law of Laplace, the magnitude of increment ΔP is

$$\Delta P = 4\gamma/R, \tag{4.1}$$

where R is the sphere radius and γ- is the surface tension equal for water at 293 K to 72.8 mN/m.

Let us use Eq. (4.1) to find the magnitude of excess pressure required to distend alveoli in inhaling. Let the value γ for the liquid lining the inner surface of alveoli equals 50 mN/m, which corresponds to the surface tension factor of intercellular liquid. Assuming $R = 0.05$ mm, we obtain $\Delta P = 4$ kPa.

Actually, Eq. (4.1) yields a twice higher value of ΔP, since the alveolar film of liquid is in contact with the air with one (inner) side only. Therefore, the true value of ΔP will be close to 2 kPa. The comparison between this value and the values of pressure required to distend a lung (Figure 4.1) makes it clear that at least a sizeable fraction of the pressure, if not the whole of it, is spent on overcoming the forces of surface tension.

Consequently, the difference between the two curves in Figure 4.1 is what contributes surface tension forces to the elasticity of a lung. In normal inhaling, lung volume roughly increases to 40–50% of their maximal volume. As Figure 4.1 shows, within this interval of the lungs distension, the contribution of surface tension forces comprises between 60 and 80%.

It's Not So Simple

Thus to a great extent, elastic recoil of a lung depends on surface tension forces. It is yet to be understood why the contribution of surface tension grows with the increase of the volume of a lung (see the exhale curves in Figure 4.1), although following Eq. (4.1), the value of ΔP should decrease with growing R. Besides, why is the contribution of surface tension dependent on a breathing phase (see the curves for exhale and inhale in Figure 4.1)? So far, we have *a priori* assumed the surface tension factor in different alveoli to be identical and independent of the state of the alveoli (distended or collapsed). Indeed, for pure liquids, surface tension factor is independent of the surface size. However, for a liquid that contains dissolved material, the molecules of the solute may accumulate spontaneously at the surface of the liquid, lowering its surface tension. Substances that act in this way are called *surfactants*.

Some surfactants have such small solubility in a liquid that once their molecules have entered the surface, they do not leave it easily. As a result, if the surface area of the liquid is decreased, the surfactant molecules get crowded together and lower the surface tension to very small values. Surface tension is then a function of surface area. If the concentration of surfactant is high and the substance can cover the entire surface of water in a continuous layer, then γ of such liquid equals γ of surfactant. When surfactant concentration is insufficient for covering the entire surface, the surface tension of the liquid will fall in between the corresponding values for water and surfactant. In these cases, the increase of the surface of the liquid will result in the reduction of the surface concentration of surfactant and in increase of γ, bringing the surface tension factor nearer to γ_{water}. It is obvious that under the reduction of the surface of the liquid the value of γ for the latter will undergo the opposite changes. Such reasoning permitted John A. Clements to explain in 1957 why the contribution of surface tension grows with the increase of the volume of a lung (for review see Clements, 1997).

In Figure 4.1, we can see that when lungs are filled with air, the plots, while coinciding at end points, have differing values in intermediate points. We obtain a so-called hysteresis in the dependence of surface tension on area, whereby, the higher the frequency of inhale-exhale cycles, the more pronounced the hysteresis is.

The lung surfactant consists primarily of phospholipids (80–90% of its mass). In addition, there are at least four distinct surfactant-associated proteins, SP-A, SP-B, SP-C, and SP-D. The phospholipids have been identified as the primary surface tension-lowering component. With the chemical composition of the lung surfactant having been established, it became possible to study the properties of a film of the surfactant that forms on the surface of water. The results of the experiment, shown in Figure 4.2, illustrate the dependence of surface tension of the surfactant film on the mean molecular area. It can be seen that, with the film shrinking, when the surface per one surfactant molecule diminishes, surface tension gradually lowers and finally levels. With increasing surface of the film, surface tension grows, closing the hysteresis loop, similar to the one in Figure 4.1.

What is the underlying cause of hysteresis in Figures 4.1 and 4.2? Why is it that, with the same surface of the film, the value of γ in inhale always exceeds that in exhale? It is related to the fact that some part of the surfactant, lowering surface tension, is situated not on the water-air interface but dissolved in deeper layers of the liquid that lines alveoli. This amount of surfactant dissolved in the bulk of liquid is in a dynamic equilibrium with surfactant molecules on the surface. However, this equilibrium is not attained instantaneously. Therefore, for instance, at the beginning of inhale, swift growth of the surface is accompanied by an abrupt increase of γ, since the surfactant dissolved in the bulk is late in coming to the surface. The equilibrium between surfactant molecules is attained only at the end of inhale (exhale), which explains the presence of hysteresis in the dependence of γ on the surface area.

Why does the surfactant have so many components? Probably the most important property of the surfactant is to lower surface tension quickly; that is, to form a stable film rapidly when the surface area is changed. Although phospholipid monolayer films can allow low surface tension, the rate of

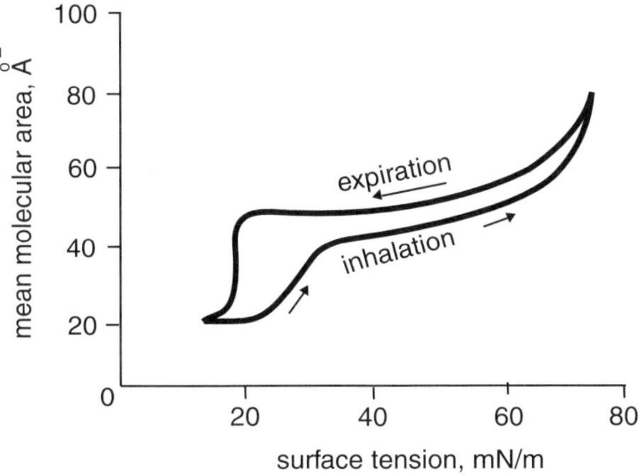

FIGURE 4.2. Surface tension, which depends on the mean molecular area for a mixed phospholipid and SP-C (0.4 mol%) film. (Modified from von Nahmen *et al.*, 1997. Reproduced by copyright permission of The Biophysical Society.)

surface adsorption to form monolayers and the spreadability of phospholipids at air-water interface are low compared to those of a natural lung surfactant (Notter *et al.*, 1980). However, as recently demonstrated in several studies (see for example, Taneva and Keough, 1994) the hydrophobic surfactant-associated proteins SP-B and SP-C enhance the surface-seeking properties of phospholipids. An advantage of having the proteins embedded in or in associated with the phospholipid matrix at low γ may be aiding in the rapid respreading and replenishment of the monolayer with the lipid upon expansion. The proteins get squeezed out of the monolayer at lower γ, but because some small amounts of proteins may remain in the monolayer at low γ they could rapidly respread the lipids from highly compressed collapsed-phase states. Although the exact nature of the lipid-protein interactions is currently unresolved, some studies suggested that interaction of long protein alpha-helices with anionic lipid headgroups might be underlie the crucial role of the proteins in the lung surfactant (Longo *et al.*, 1993).

Where do the substances that lower surface tension and thereby facilitate breathing in a lung come from? It turned out that they are synthesized by special cells in alveoli. The synthesis of these surfactants is underway for the whole life of a man, from his birth till his death. In those rare cases when in a newborn's lungs there are no cells producing surfactants (respiratory distress syndrome), a baby is unable to take its first breath by itself and dies. Unfortunately, about half a million newborn children all over the world die every year without taking their first breath because of a shortage or absence of a lung surfactant in their alveoli. Shortly after the lung surfactant was isolated and characterized, attempts were made to treat respiratory distress syndrome of a newborn with aerosols of this substance. The treatment is now standard around the world (Poulain and Clements, 1995).

Exceptions to the Rule

However, many animals that breathe with lungs feel perfectly comfortable without any surfactants in their alveoli, for example, the cold-blooded—frogs, lizards, snakes, crocodiles. Since these animals don't have to spend energy on heating their bodies, their oxygen demand is roughly an order of magnitude lower than that of warm-blooded animals. This is exactly why the lung area used for gas exchange between blood and air is smaller in cold-blooded animals than in warm-blooded. Thus one cm^3 of air in the lungs of a frog has a mere $20\,cm^2$ surface of contact with blood vessels, whereas in man the same air volume exchanges gases with blood through the surface of about $300\,cm^2$.

The relative decrease of lung area per unit of its volume in cold-blooded animals is due to the fact that diameter of their alveoli is roughly 10 times that of warm-blooded animals. Notice that the law of Laplace implies that the contribution of surface tension forces is in inverse proportion to the alveoli radius. Therefore, the large radius of alveoli in cold-blooded animals renders their distension easy even in the absence of surfactant on their inner surface.

Birds are another group of animals with no surfactants in their lungs. Birds are warm-blooded animals and lead rather active lives. Energy outlays of birds and mammals of the same weight are similar, as is their demand of oxygen.

A bird's lungs possess a unique capability of saturating blood with oxygen in flight at a great height (about 6000 m), where its concentration is half that at sea level. Any mammals (including man), once at such a height, begin to experience lack of oxygen and radically scale down their motional activity, sometimes lapsing into a subconscious (comatose) state. How do the lungs of a bird, without a surfactant, manage to breathe and saturate blood with oxygen better than the lungs of mammals?

Let us engage in some self-criticism. What is wrong with our lungs? First, not all the inhaled air takes part in a gas exchange with blood. Namely, the air left in trachea and bronchi at the end of inhaling is unable to give oxygen to blood and take carbon dioxide out of it since there are almost no blood vessels there. For this reason, the part of the lungs' volume taken up by trachea and bronchi (along with the volume of the upper respiratory tract) is usually called *dead space*.

Typically, the dead space in human lungs has a volume of about 150 cm^3. Notice that the presence of this space not only prevents a corresponding amount of fresh air from accessing the inner surface of alveoli, abundant in blood vessels, but also reduces the average oxygen concentration in the part of air that has reached alveoli. It occurs due to the fact that at the start of each inhaling the alveoli are filled with the dead space air, which is the last portion of the air that has just been exhaled. Therefore, the oxygen concentration in the air that enters alveoli at the start of inhaling is low, and does not differ from that in exhaled air.

We can artificially increase the dead space volume by breathing through a long pipe. You will likely notice that the depth (volume) of breathing in this case must be increased. Evidently, if a dead space volume is made equal to the maximal possible inspiration (i.e., about 4.5 L), a person will start gasping in several breaths as there will be no fresh air at all to enter the alveoli. Thus, the existence of dead space in the respiratory system of mammals is clearly a miscalculation on the part of nature.

While creating the lungs of mammals, regrettably, nature made another mistake—the air motion in lungs changes direction in its transition from inhaling to exhaling. Therefore, roughly half of the time the lungs are practically idle since no fresh air enters the alveoli during the exhaling phase. As a result, at the end of exhaling the oxygen concentration in the alveolar air diminishes one and a half times compared to that in the atmosphere. Since during inhaling the oxygen-rich inhaled air mixes in the alveoli with the air already there, the resulting mixture, which will exchange gases with blood, contains oxygen in a smaller concentration than that in the atmosphere. Therefore, blood oxygenation in mammals will always be less intense than in a hypothetical case where the air would pass through the lungs all the time in the same direction regardless of the breathing phase.

Of course, in the lungs of mammals with the trachea serving for both air entry and exit at once, this unidirectional motion of breathing mixture is unfeasible. Meanwhile, in birds nature again reached perfection. Besides normal lungs, birds have an additional system composed of five or more pairs of airbags connected to lungs. The cavities of the bags have many branches in the body, entering some bones, sometimes even the fine ones. As a result, the respiratory system in a duck takes up about 20% of body volume (2% lungs and 18% air bags), whereas in a human, a mere 5%. Airbags not only reduce the body density but also promote the blowing of air through the lungs in a single direction.

The only thing that changes in birds' breathing is the volume of airbags, whereas the volume of a lung remains practically constant. Because there is no need to expand the lung, it becomes immediately clear why there is no surfactant in birds' lungs — it would just be of no use there.

Countercurrent: Cheap and Effective

Seeking to maximally increase the oxygen concentration in the blood of birds during their flights at great heights, nature resorted to yet another contrivance: the direction of blood motion in birds' lung vessels is the opposite of that of the air current through a lung. Such countercurrent manner of blood oxygenation is much more efficient as compared to the case when the blood and the air move through lungs in the same direction. Let us demonstrate it with the following example.

Assume that two tubes, which simulate the adjacent blood vessel and the air-carrying tube of a bird's lung are in contact with each other along a certain segment (Figure 4.3). The surface of contact between the blood vessel and the air-carrying tube enables oxygen to diffuse from air into blood, with carbon dioxide diffusing in the reverse direction.

The blood that is about to leave the lung (the right-hand part of Figure 4.3) is in contact with the air that has just entered the lung, in which oxygen concentration has not yet been lowered. As the air passes through the lungs, it loses oxygen and takes in carbon dioxide. Therefore, moving along the vessel, blood comes in contact with the increasingly oxygen-rich portions of

FIGURE 4.3. Countercurrent method of blood oxygenation in bird lungs.

fresh air, which provides it with the opportunity to be oxygenated to the greatest possible extent. The same mechanism enables blood to get rid of excessive carbon dioxide faster than it happens in mammals.

Interestingly, nature employed the countercurrent system not only in birds, which encounter lack of oxygen in the flights at great heights, but also in the gills of fish, which use oxygen dissolved in water, where its concentration is roughly 1/30 of that in the atmosphere.

Diving

In man and other higher-organized animals, breathing and the beat of the heart are synonymous to life. Heart and lungs supply an animal with the required amount of energy, deliver oxygen to tissues, and remove carbon dioxide from them. For this reason cessation of breathing or of blood circulation represents a great danger for the life of an animal.

However, not all tissues need uninterrupted oxygen supply to the same degree. If blood circulation in a hand or a leg is stopped with a tourniquet for an hour or even longer, it will cause no damages to the tissues of these organs. Kidneys can also endure interruptions in blood supply. Unfortunately, the heart and the brain are highly sensitive to the lack of oxygen. Therefore, asphyxia or cardiac arrest that lasts several minutes results in irreversible changes in the respective tissues.

It is known that cats, dogs, rabbits, and other land mammals die several minutes after they have been completely submerged in water. Ducks, however, can endure 10–20 minutes of submersion underwater; seals, 20 minutes and longer; some species of whales, more than an hour. How can they do it?

Experiments performed on seals have shown that during diving the cardiac rhythm in these animals sharply slows down (to 1/10 of normal rate). This occurs right after submersion of their nose openings underwater. The same event takes place in penguins, crocodiles, tortoises, ducks, and all other animals who breath air, but spend part of their time underwater. It is of interest here that in flying fish, whose gills stop working when the fish leap out of water or are forced out of it, heart contractions noticeably slow down as well. Such sharp deceleration of the cardiac rhythm under conditions of oxygen deficiency in all these animals enables a radical reduction of oxygen consumption by the heart, which is the main oxygen user in an organism (for review see Butler and Jones, 1997).

In diving, for the blood supply of the heart and brain not to fall below the permissible level, the diameter of vessels in other organs (except the heart and brain) decreases noticeably. Therefore, even at a slow rate of heart contractions, the oxygen supply of the heart and the brain in diving animals is still sufficient. The same regulating mechanism of blood circulation in diving is developed in training pearl divers, who are known to be able to stay underwater for several minutes, at a depth of up to 30 m.

What about us, ordinary people, who do not have the capabilities of pearl divers? How can a common man investigate the mysteries of the sea depths? The world's ocean, with the average depth of about 3 km and the area comprising 70% of the planet's surface, is still practically unexplored. And, though in 1960 the bathyscaph *Trieste* descended 11 km to the deepest part of the ocean, at present there are fewer footmarks left by man even at a depth of 1 km than on the surface of the moon.

The first contraption for a continued stay of man underwater seems to have been a long pipe connecting his mouth to the atmosphere. A breathing pipe was already used by ancient Greeks and Romans. Leonardo da Vinci (1452–1519) improved a breathing pipe, supplying it with a cork disk placed so that the upper end of the pipe always extended above the water and a person could breathe freely. The length of the pipe was up to a meter. Leonardo da Vinci's pipe was intended not for underwater swimming but rather for "walking under the water." The great scholar thought that the contraption could be used in the Indian Ocean to "mine for pearls." It is of interest that a breathing pipe of a kind is found in larvae of some insects living at the bottom of puddles and shallow ponds. By thrusting their pipe up to the surface of water, they have the possibility to breathe without getting out of mud.

Judging by James F. Cooper's novels, the Indians often resorted to a breathing pipe: as they were hiding from their enemies under the surface of water, they were breathing through a hollow reed. However, this method of breathing underwater is possible to use only when the depth of submersion does not exceed 1.5 m. At a greater depth of submersion, the difference between the water pressure compressing the chest and the air pressure inside it increases so much that we are already unable to increase the chest volume in inhaling and fill the lungs with fresh air. Therefore, while at a depth greater than 1.5 m, we can only breathe the air compressed to the pressure equal to that of water at this depth. To this end, skin divers take along compressed air bottles. However, descent to different depths requires different pressure of inhaled air. Thus, at a 10 m depth the pressure must equal 200 kPa, and at a 40m depth, –500 kPa. Therefore, a skin diver should continually monitor the depth of descent and change the inhaled air pressure accordingly.

Unfortunately, the experience of using Aqualungs has shown that they can be used to descend to a depth less than 40m. At a greater depth, a skin diver would have to breathe air that would be compressed to a pressure in excess of 500 kPa, and would have an oxygen concentration more than five times that of the atmosphere, which would cause oxygen poisoning.

A person can breathe pure oxygen at the atmospheric pressure for about 24 hours only. A longer period of breathing oxygen brings on pneumonia, terminating in death. A person can breathe pure oxygen compressed to 200–300 kPa for one and a half to two hours at most. Afterwards, he or she suffers disturbances of coordination of movements, as well as attention and memory disturbances.

To preclude the toxic effect of oxygen, skin divers who descend to a great depth are supplied with special breathing mixtures, where the oxygen percentage is lower than that in the atmospheric air. However, at such high pressures, the nitrogen in the breathing mixture can produce a narcotic effect.

In addition, breathing nitrogen-bearing mixtures at a depth of about 100 m is very hard, as the density of inhaled gas compressed to the pressure of 1000 kPa is ten times that of the atmospheric air. Such a high density of inhaled gas transforms an otherwise easily performed act of breathing into a process of the labored "pushing" of air into the lungs. Therefore, as a rule, at depths below 40 m divers breathe a mixture of oxygen and helium. Helium has no narcotic effect at such high pressures and its density is approximately 1/7 that of nitrogen.

Yet divers descend deeper and deeper. They often have to mount and replace oil rigs at sea where oil is the cheapest since it occurs at a shallow depth. The deep-sea divers who mount rigs in the North Sea at times have to work at a depth of about 300 m and breathe gas mixtures compressed to a pressure of 3000 kPa.

Nonetheless, hardships (and even dangers) lie in wait for a skin diver that has descended to a great depth not only underwater, but also right after he or she comes up to the surface. Long ago it was already known that deep-sea and skin divers who quickly ascend from a great depth soon begin to experience intense pain in their joints. This occupational disease of divers became known as decompression illness. It turned out that the unpleasant sensations in the joints of divers who had just ascended from depth were due to formation of gas bubbles in tissues. The gas bubbles can also be the cause of clogging of small blood vessels.

In its 1997 edition of *Report on Decompression Illness and Diving Fatalities*, Divers Alert Network states that the number of reported cases of decompression illness that occurred in the diving year 1995 was 1132, including 104 fatalities from recreational dives.

Decompression sickness results from gas coming out of the solution in the bodily fluids and tissues when a diver ascends too quickly. This occurs because decreasing pressure lowers the solubility of gas in liquid. Rapid ascent may lead to bubble formation. The bubbles emerge in the same way as they do in a soda water bottle once it is opened. In both cases, the bubbles arise under lowering of pressure above the liquid saturated with gas at high pressure. Henry's law and Dalton's law are central to understanding decompression sickness. Henry's law states that, at a given temperature, the amount of gas that will dissolve in a liquid is directly proportionate to the partial pressure of the gas. Dalton's law states that the pressure of a gas is the sum of the partial pressures of all gases present.

Decompression illness is also possible at a quick climb to a height in an unsealed chamber. In this case, the danger of decompression illness arises due to a sharp drop in pressure roughly by 50 kPa (at a height greater than 6000 m). Several cases of decompression illness have been recorded in pilots flying in an unsealed cabin at an altitude of about 2500 m. However, on the day preceding the flight all these persons went skin diving with the use of an Aqualung. Obviously, before the flight the organism of each pilot contained small air bubbles, which started expanding and were perceived by the pilots upon an insignificant decrease of the atmospheric pressure. Therefore, pilots are recommended to get behind a steering wheel of an airplane at least two hours after skin diving.

For a bubble to be formed at the site where it has not been before, its evolution should, obviously, pass through two different phases: 1) the formation of the tiniest bubble at the site where "there was nothing," and 2) bubble growth. Gas bubble growth at a sudden drop of atmospheric pressure is easily explicable with the help of Boyle's law.

The mechanism of the tiniest gas bubble formation "out of nothing" has so far been studied insufficiently. It is believed that under normal conditions there always are so-called micronuclei present in body tissues. Those may be the precursors of decompression illness bubbles. The presence of micronuclei appears to be necessary for the bubble formation process, as in pure water, when gas bubbles fail to form at all even at a sudden thousandfold drop of gas pressure above its surface. Possibly such nuclei might include stable (retaining their size) gas bubbles present in tissues, whose selected mechanisms of stabilization were examined by van Liew and Raychaudhuri (1997). They reviewed the mechanisms that can overcome the absorptive tendencies, so that small spherical bubbles can persist. One general type of stabilizer is a mechanical structure at the gas-liquid interface that can support a negative pressure so that gases inside can be in diffusion equilibrium with their counterparts outside. One possibility for such a mechanical stabilizer is surfactant films.

Decompression illness can be avoided if a diver is lifted from a great depth slowly enough, with necessary pauses. Such pauses in the ascent enable dissolved gas to diffuse through tissue to blood vessels. From there it passes into lungs along with the blood flow, and then further into the atmosphere, having failed to form bubbles. It is believed that decompression illness does not develop at a sudden ascent from a depth less than 9 m.

To reduce the decompression illness risk as much as possible, the diver has to adhere to one of the safe diving practices and select an appropriate dive profile and decompression table. For example, according to the National Association of Underwater Instructors, the stay at a 24 m depth for an hour makes it necessary to stop during the ascent at a depth of 5 m for 17 minutes.

In the cases when the divers work daily at depths in excess of 100 m for all working hours, it has been considered expedient that the pressure of the air they inhale not be lowered even in a recreation period after the ascent from the depth, as it would require several hours. Therefore, in the interval between descents they rest in special hyperbaric chambers installed on board ships.

The greater part of gas bubbles is formed by nitrogen, for oxygen is intensely consumed by body cells. The danger of decompression illness development can be reduced if instead of nitrogen helium is used, which is 36% less soluble in blood plasma and 76% less soluble in fats. Besides, the diffusivity of helium through body tissues is almost three times that of nitrogen. Great diffusivity of helium permits a shorter duration of a diver's ascent to the surface.

It turned out, however, that helium-oxygen mixtures ensure the normal work of divers only to a depth of 400–450 m. At further increase of pressure, the mixture density becomes very high, which renders breathing impossible. Much is expected of the inclusion of hydrogen—the lightest gas in the mixture for divers. Naturally, a gas mixture containing hydrogen and oxygen simultaneously is explosive. However, the probability of mixture explosion is very small, as the ratio of oxygen and hydrogen volumes is far from the explosive one (1:2). At the same time, on the surface, where the gases are stored and mixed, the probability of explosion is much greater, which requires the undertaking of necessary precautions.

To explore the relative dangers of different inert gases, the effects of physical properties of the gas on decompression-sickness bubbles were studied using a mathematical model (Burkard and van Liew, 1995). However, the task of finding the optimal gas mixture, which could let deep-sea divers master depths exceeding 500 m, is still unresolved.

There is yet another phenomenon related to the formation of gas bubbles in liquids at a sudden lowering of external pressure. In our nervous age, it is rather common to crack our fingers in minutes of agitation. For a very long time, the cause of the cracking sound when pulling joints has remained unknown. Many believed that it was the clicking of bones. Upon a detailed study, however, it turned out that the cause of these cracking sounds was gas bubbles that formed and burst in the liquid filling the sinovial capsule. When a joint is pulled, the volume of the sinovial capsule increases, the pressure in it drops accordingly, and the fluid lubricant in the joint boils up. The forming gas bubbles merge with the larger ones and burst with a pop. When the bones return back to normal position, the gas is gradually absorbed by the liquid. This takes place for about 15 minutes. After this time, it is possible to crackle the joint again.

All of these difficulties related to the stay of man underwater arise because compressed air is breathed. What if a person is made to "breathe" water as fish do? Of course, oxygen concentration in water at equilibrium with the atmosphere amounts to less than 5 percent of that in the air. But even this concentration, when in contact with blood, will be enough to saturate the latter with oxygen to a normal level. In addition, the oxygen concentration in "inhaled" water can be increased by continuously passing pure oxygen through it rather than air.

Obviously, in "breathing" water containing dissolved oxygen, there is already no need to compensate for the increase of external pressure in descent since according to Pascal's law the pressure of water inside lungs will always equal the outside pressure. Therefore, the efforts required to inhale will not change with the depth of descent. Using water as the carrier of dissolved oxygen saves one from the danger of oxygen poisoning, for the oxygen concentration in "inhaled" water can be made constant and equal to that in the atmosphere. For the same reason, the danger of decompression illness does not exist anymore.

With the help of a special apparatus, dogs and mice were able to live "breathing" water for several hours. They died because the concentration of carbon dioxide in their blood grew beyond the permissible limit. Thus, "breathing" water, animals completely met their demand in oxygen but were unable to remove the forming carbon dioxide efficiently.

In mammals under normal conditions (at rest), each liter of exhaled air contains about 50 mL of CO_2, whereas the solubility of this gas in water is such that each liter of it under the same conditions can contain no more than 30 mL of CO_2. Therefore, to remove all the carbon dioxide forming in an organism, it is necessary to pump through the lungs almost twice as large volumes of water as the required volumes of air.

According to Bernoulli's equation, the difference of pressures needed for moving liquid (or gas) medium through a pipe of the known length and diameter at a certain velocity, must be proportionate to the viscosity of the medium. And as the viscosity of water is roughly 30 times that of the air, unaided "breathing" of water will require approximately 60-fold energy outlay. Since nature endowed us with the lungs that are useless in sea depths, to explore these depths we need bathyscaphs and submarines. But is everything really that hopeless?

In 1966, Clark and Gollan established the remarkable gas exchange qualities of perfluorocarbons (PFC) — liquids that are structurally similar to hydrocarbons, with the hydrogen replaced by fluorine. Their experiment involved submersing spontaneously breathing mice in the PFC. The animals sustained life while immersed in the liquid and survived after their return to gas breathing. This experiment was the breakthrough required for further investigation of the use of PFC for support of gas exchange in the lung (for review see Sadowski, 1996).

PFC liquids have many qualities that are unique and that make them perfect for liquid ventilation. PFC liquids are nontoxic, odorless, and aren't absorbed by tissues. PFC liquids have 16 times the oxygen solubility and 3 times the carbon dioxide solubility of water. This gives PFC liquids an excellent oxygen and carbon dioxide carrying capacity: 50 mL of oxygen per dL and 160–210 mL of carbon dioxide per dL. PFC liquids have a slightly smaller carrying capacity of oxygen compared to that of blood: up to 200 mm Hg. But beyond 200 mm Hg, PFC liquids are actually better.

Another trait that makes PFC liquids suitable for liquid ventilation is a very low surface tension (approximately 18–19 mN/m). When instilled, the PFC spreads into the collapsed alveoli where gases have been unable to penetrate before. It lines the alveolus, stabilizing and re-expanding it. PFC liquids are twice as dense as water; therefore, liquid ventilation requires the use of a mechanical breathing device to assist breathing.

More recent research in animals and limited studies in humans have shown that oxygen-carrying fluorocarbons have the potential to improve pulmonary function in both infant and adult respiratory distress syndromes. Further biophysical studies are needed to optimize delivery of the fluorocarbons and their adjunctive use with mechanical ventilation.

High Frequency Ventilation and Einstein's Formula

The techniques of resuscitation of organisms must have always interested people. Apparently, the earliest of survived methods of mechanical ventilation is contained in the book *De humani corporis fabrica* (1543) by the medieval Flemish anatomist A. Vesalius (1514–1564), a founding father of anatomy. At those times, reviving departed souls was considered a seditious business, so Vesalius, persecuted by the church, explains in one of his books how to apply mechanical ventilation to pigs. His views on functions of various bodily organs were condemned as heresy and he had to undertake a pilgrimage to Palestine, never to return, dying in a shipwreck.

It is interesting that in the same year, Copernicus published his historic treatise *De revolutionibus Orbium Caelestium* that made a breakthrough in natural science. The same year provided mankind with a possibility to change its views radically on man and the universe. But this failed to happen, and both geniuses were declared heretics. However, the progress of knowledge cannot be stopped, and a century later (1667) Robert Hooke (1635–1703), known to many of us as a physicist, reproduced almost word for word the method of mechanical ventilation described by Vesalius.

Many years have since passed, and today the technique of mechanical ventilation enjoys wide application in clinics during surgery and in first aid for reanimation of seriously ill patients. Modern lung ventilation units present a pump to force breathing mixtures (for severe cases, pure oxygen) into the patients' trachea. The pump has special valves, V_1 and V_2 (Figure 4.4); the first is for connection of the pump and lungs in inhaling and disconnection in exhaling, and the second is for connection (during exhaling) and disconnection (during inhaling) of trachea and the atmosphere.

Are physicians happy with such an artificial lung? Not always. Patients often develop post-surgical complications such as pneumonia. This is attributed to the rupturing of the alveoli during mechanical ventilation. But why do the alveoli not rupture when we breathe on our own?

As we know, lungs are enclosed on all sides with an airtight pleural cavity filled with air, the latter connecting the former mechanically with the thorax. In unaided breathing, the muscles of thorax, contracting at the beginning of the phase of inhaling, cause its volume to increase so that the pressure inside the lungs drops below the atmospheric one, and a fresh portion of air enters the lungs through the trachea. Therefore, it is obvious that in this unaided inhaling, the pressure gradient between the inner and outer cavities of each alveolus will be determined solely by its elastic performance and the characteristics of surfactants.

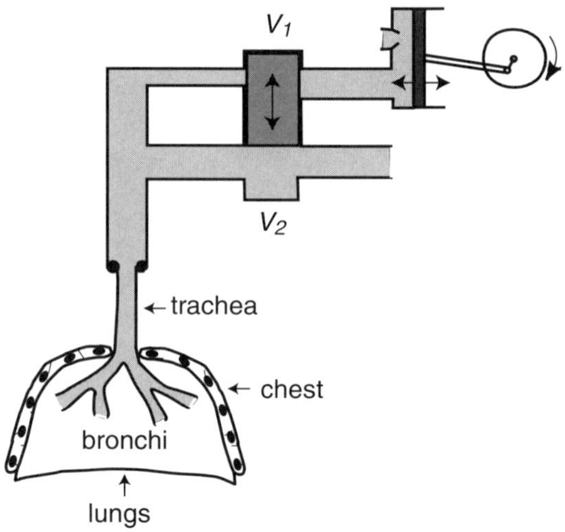

FIGURE 4.4. Schematic representation of conventional mechanical ventilation. V_1 and V_2 are the valves changing the direction of air flow through the lungs.

Under mechanical ventilation, the air pressure inside an alveolus in inhaling should not only insure its adequate distension but also increase the volume of the entire thorax. Thus, under mechanical ventilation, the thorax enclosing the lungs represents a substantial obstacle that entails the significant increase of the the air pressure in inhaling.

The measurements have shown that in unaided breathing with a frequency of 14 inhales a minute and a volume of 0.5 L, the excess pressure inside alveoli never exceeds 0.1 kPa, whereas under mechanical ventilation it may be as high as 0.5–1.0 kPa. The result of the growth of air pressure in alveoli is their ruptures and subsequent complications. What can be done to make mechanical ventilation safer?

The solution is self-evident: the tidal volume must be decreased and, at the same time, the breathing frequency must be raised. Indeed, by diminishing the distension of lungs, we decrease the magnitude of the excess pressure in them, and, consequently, the probability of their rupture. However, to use this method of optimization of mechanical ventilation, one has to know the interrelation of breathing frequency and the tidal volume.

It has been established that every minute, 5 L of *fresh* air must enter a person's lungs in a state of rest. The word "fresh" is highlighted here to remind you of the existence of dead space, from where no fresh air enters the lungs. Thus, the following relation between breathing frequency f[1/min], tidal volume V_t[L], and dead space volume V_d holds

$$5 = (V_t - V_d)f \tag{4.2}$$

To make use of Eq. (4.2), we have to know the numeric value of V_d. We already know that the average V_d of an adult is about 150 mL, but how is the value of V_d to be found for every one of us?

In order to do it, it is sufficient for a person to take a single breath of pure oxygen rather than air from the atmosphere. Then the gas composition of exhaled air has to be studied. The analysis of this first exhalation after inhaling pure oxygen is shown in Figure 4.5.

The first 100 mL of exhaled air show no trace of nitrogen, which means that this air did not mix with the air in the lungs. Therefore, the volume of dead space in this case can be considered equal to 100 mL. Subsequent portions of exhaled air do contain nitrogen, bearing evidence of the gas exchange taken place between the inhaled O_2 and the N_2 present in the lungs.

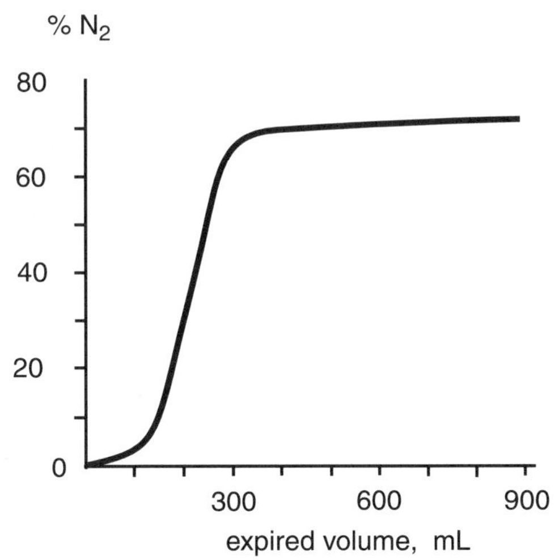

FIGURE 4.5. Single-breath N_2 curve. (Modified from Permutt *et al.*, 1985. Reproduced by copyright permission of The American Physiological Society.)

When the validity of Eq. (4.2) came under scrutiny, it proved to be valid only for $V_t > 0.4\,\mathrm{L}$, while yielding too-high values of f at smaller V_t. Pulmonary and cardiac surgeons came to dislike the formula also because it *forbade* using $V_t < 100 - 150\,\mathrm{mL}$ in mechanical ventilation, because movements of lungs while breathing get in their way as they operate.

And so, in the late 1960s, several surgeons at once, forgetting Eq. (4.2) and dead space, decided to cross the forbidden line, let V_t go below 100 mL and . . . succeed. Their patients endured a seemingly inexplicable type of mechanical ventilation perfectly well. Thus, for instance, mechanical ventilation with $V_t = 50\,\mathrm{mL}$ and frequency $180\,\mathrm{min}^{-1}$ reliably secured vital functions of a human organism at rest. If we substitute the numbers in Eq. (4.2) and evaluate V_d, it turns out that it should equal a mere 20 mL. What is V_d equal to, after all? 150 mL or 20 mL?

Before answering this question, let us try to answer another one. What role in breathing does diffusion play during the movement of air from the atmosphere into lungs (and back)? Obviously, a very insignificant one. Indeed, it is sufficient to hold your breath for one minute (even after the preliminary breathing of pure oxygen), and an irresistible desire to inhale fresh air arises.

What is the way to enlarge the role of diffusion? In the middle of the 1960s the same problem was faced by the scientists dealing with physics of soil, when they set a task to oxygenate deeper layers of earth. Their research, performed on different porous materials, showed that the gas diffusion through these media accelerates significantly in the presence of oscillatory motion of air. Thus, the index of oxygen diffusion through a test container with a porous medium increased four times (compared to standard conditions), whereby a mere 1/200 of volume under study was pumped through it (there and back) at a frequency of 4 Hz.

According to Fick's law, gas diffusion coefficient, D, through surface S of a layer of thickness x, can be found from the following relation:

$$D = (V)'[x/S(C_1 - C_2)] \tag{4.3}$$

where $(V)'$ is the velocity of the gas diffusion current through the layer, and C_1 and C_2 are the concentrations of the gas on both sides of the layer. This equation can, obviously, also be used for measuring the diffusion currents between the lungs and the atmosphere. As a physical model of space located between lungs and the atmosphere, Jaeger *et al.* (1984) used a pipe of 60 cm in length and 2 cm^2 cross-section area, which corresponds to the average cross-section area of trachea and bronchi and their length.

With the help of the model, Jaeger *et al.* studied the dependence of D on frequency f of longitudinal vibrations of air in the pipe and the amplitude V_t (in mL) of the vibrations. Figure 4.6 shows how D increases with growing f and V_t. It turned out that D is in direct proportion to f and $(V_t)^2$. In addition, direct proportional dependence of D on S^{-2} was revealed in special experiments. Thus, the empirically found dependence of D on these parameters was of the form

$$D = kf(V_t/S)^2 \tag{4.4}$$

where k is a dimensionless factor, close to 0.05, given f is measured in Hz, S in cm^2, and V_t in mL.

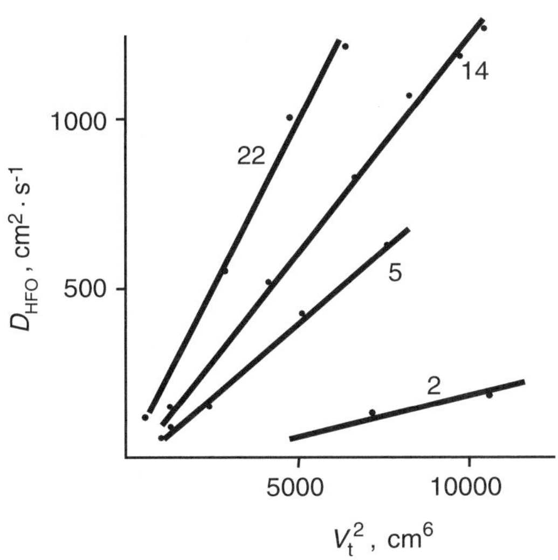

FIGURE 4.6. Results from a bench model. D_{HFO} values, computed with Eq. (4.3), are plotted against the square of tidal volume, $(V_t)^2$. Frequencies were 2, 5, 14, or 22 Hz and marked near corresponding lines. (Modified from Jaeger *et al.*, 1984. Reproduced by copyright permission of Lippincott, Williams & Wilkins.)

It follows from Eq. (4.4) that gas diffusion coefficient in a long pipe is proportionate to the frequency of vibrations of gas particles and to the squared amplitude of the vibrations. This dependence is very much like the relation derived by Einstein in 1905 for molecular gas diffusion. The diffusion coefficient in Einstein's formula is directly proportionate to the collision frequency of molecules and to the squared length of their free path. Therefore, Eq. (4.4) will coincide with Einstein's formula, if the frequency of vibrations of air column in the pipe is put in correspondence with the collision frequency of gas molecules, and the amplitude of longitudinal vibrations of the column, with the free path length of a molecule. Thus, diffusion and convection currents, arising in a pipe filled with gas, can increase the diffusion coefficient of oxygen by a thousand times.[1] This explains the efficiency of high-frequency oscillatory ventilation.

Still there remain some patients who need very low ventilation pressures despite the most sophisticated conventional ventilation strategy. One of the alternatives for these is the high frequency ventilation that differs from conventional positive pressure ventilation in rates greater than 50 breaths/min and a tidal volume similar or less than dead space. Another alternative is high frequency jet ventilation that uses a pulse of a small jet of fresh gas introduced from a high-pressure source (50 psi) into the airway via a small catheter. The rates here are usually 100–400 breaths/min.

High frequency ventilation can be utilized with all patient populations suffering from acute lung injury. Unfortunately, so far no complete theory of the high frequency ventilation has been proposed, which somewhat delays the application of this efficient technique in clinics.

The Physics of Cough

Cough is known to arise either as a result of the infectious diseases of the respiratory tract or because of foreign bodies or small particles that have gotten into it. In all these cases, cough serves for discharging mucus and particles that obstruct normal breathing. Evidently, the greater the linear velocity of air escaping trachea during cough, the more effective the cough

[1] Note, for comparison, that D for oxygen diffusion in nitrogen atmosphere with no oscillations amounts to about 0.2 cm^2/sec at room temperature.

is. Let us try to evaluate the maximal magnitude of the velocity and sort out on what parameters of the respiratory system it depends.

Before coughing, we inhale air into lungs. After that, following an inborn reflex, the glottis is closed (i. e., trachea no longer communicates with the atmosphere), while the contracting muscles of thorax begin to decrease its volume. As a result (as a rule), the lungs' volume diminishes from 6 to 5 L, whereas the air pressure in the lungs increases by 20 kPa. This phase of compressing the air in lungs lasts for about 0.2 sec. Then muscles of the glottis suddenly relax and the compressed air from the lungs escapes outside with a distinctive sound.

Figure 4.7 shows how the volume rate of air outflow from lungs changes in time during coughing. The maximal volume rate in this case roughly equals 6 L/sec; however, sometimes it can become as high as 10–12 L/sec, which should be considered its maximum. What leaps to the eye is that the volume rate spike lasts for a very short time (less than 0.1 sec), after which the rate falls off sharply. As Figure 4.7 shows, arising simultaneously with the end of the volume rate spike is the "sound track" of cough. What hinders sustaining the high velocity of air effluence from the trachea for a longer time, and what is the cause of the sound?

The answer to the first question would seem to be obvious. A person's thoracic muscles are unable to secure prolonged and sufficiently high compression of air in the lungs. Indeed, the physical capabilities of a person are limited. But let us attempt to simulate the process of air effluence from trachea. Instead of a human trachea, we can examine a trachea of a large animal (such as a dog) or pick an elastic pipe with similar mechanical parameters.

Figure 4.8 shows the volume rate of air outflow from such a pipe as a function of the difference of pressures between its ends. The volume rate of air outflow proves to grow with increasing pressure, but it has an upper bound, so that with the pressure beyond 5 kPa the growth of the volume rate stops, whereas at further increase of the pressure the volume rate even starts falling. Strange, isn't it? What about Hagen-Poiseuille's equation? And more generally, where does the energy go? Why does the growing pressure with which we increasingly compress the air in the pipe fail to transform into the kinetic energy of the motion of air particles at the exit from the pipe?

The answer is simple. Staring at the plot and speculating quite often proves to be insufficient to explain a physical phenomenon. This is what

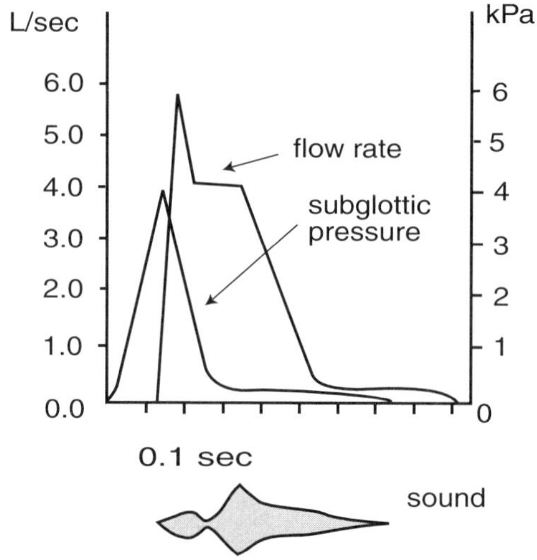

FIGURE 4.7. Flow at airway opening (flow rate), subglottal pressure, and sound level during a representative cough. (Recordings are modified from Yanagihara *et al.* 1966. Reproduced by copyright permission of Scandinavian University Press.)

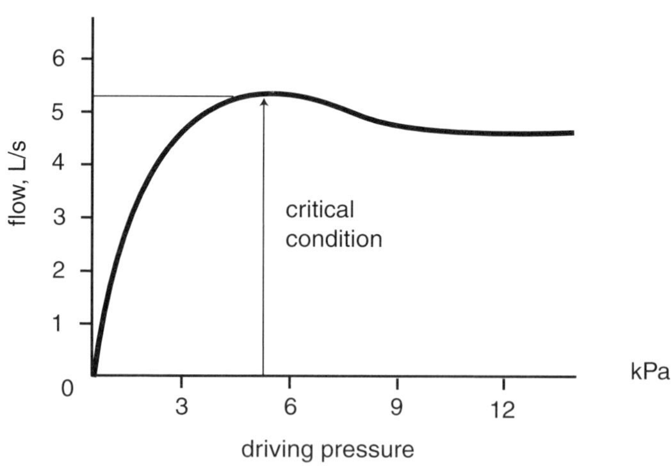

FIGURE 4.8. Progressive decreases in pressure at the downstream end of excised dog trachea result in increasing flows only to critical condition (arrow). (Modified from Elliott and Dawson, 1977. Reproduced by copyright permission of the American Physiological Society.)

happens: with the volume rate of air outflow from the pipe attaining its maximal values, its elastic walls begin to experience compression, now closing, now opening, the passage in the pipe for the air to move. As a result of such oscillations, the average registered velocity of gas outflow ceases to grow with the increase of pressure, and the energy of compressed air transforms into the energy of the strain of pipe walls and, of course, into the heat. Also, we have no reason to question the validity of Hagen-Poiseuille's equation, it being valid for pipes with rigid walls only.

Obviously, it is the periodic changes of the velocity of air outflow from trachea that cause the appearance of the distinctive sound in cough. What underlies these oscillations of the velocity of outflow air in cough? This is not easy to answer. First, let us try to answer the following seemingly silly question: Why should the air particles in the middle of the pipe increase their linear velocity, when the air pressure at one of its ends grows? How do they know of the pressure growth?

Of course, from neighbors; and those, in turn, from their neighbors, and so on. In other words, if pressure at one of the pipe ends increases (or reduces), a compression wave travels along the pipe with a *certain* velocity, thus "informing" of the change in pressure. And as soon as the wave reaches a specific air particle in the middle of the pipe, the particle increases (or reduces) its velocity.

What determines the magnitude of the propagation velocity of compression wave? If you have read Chapter 2, "Heart Pulse," perhaps you have already guessed that the velocity of a compression wave that travels along the trachea is determined by the same parameters as velocity θ of the pulse wave that propagates along a blood vessel (see Eq. (2.1)); that is, by the trachea diameter, its wall thickness, Young's modulus of the tissue it consists of, and the density of air.

Now let us imagine that the velocity of air motion along the trachea exceeds θ. How, indeed, would the air particles know that they have to increase (or reduce) the velocity of motion since the compression wave, which used to inform them of it, will never catch up with them? Therefore, it follows that the velocity of air motion along the trachea cannot exceed θ. The measurements of the maximal velocity of air outflow in coughing have shown that the value in question is, typically, below 100–150 m/sec, which agrees with the estimates obtained using Eq. (2.1).

There is yet another question to answer: Why do the pipes come under compression when the air is passed through them? When we blow air of density ρ at velocity ν through a pipe, pressure *(P)* inside it can be evaluated using Bernoulli's equation; so, if the losses for viscous friction are ignored, we obtain

$$P = P_O - \rho v^2/2 \qquad (4.5)$$

where P_O is the atmospheric pressure. Thus, air pressure inside trachea during exhaling and coughing can be less than the atmospheric one; this means the trachea is acted upon by the forces of contraction, which seek to lessen its cross section.

One more thing: the properties of a real pipe or of a trachea are not uniform across its length. This means that there is a point in the pipe where the rigidity of its walls is minimal. It is at this point that the trachea opening becomes the smallest possible, with the linear velocity of the air motion increasing accordingly. As Eq. (4.5) shows, this results in an even greater contraction along this section, and so on, until the linear velocity of gas at the section reaches θ.

Obviously, when the trachea opening at the least rigid section diminishes noticeably, the loss for viscous friction grows significantly, and cannot be ignored any more. Therefore, the relationship between P and ν in this case will not observe Eq. (4.5).

The flow of liquids and gases along pipes with elastic walls is a very complicated process. Giving its exhaustive description with the help of vivid and easy-to-grasp examples the way we try to do here, is practically impossible and, perhaps, should not be even attempted. For this reason, I want to place a period at this point—practically in mid-sentence. Those of you who have gotten really interested in the physics of cough, with many questions still unresolved, are referred to the recently published work of Walsh *et al.* (1995).

5

Hunt for Cells in an Electric Field

We are aware of the fact that we consist of cells, but we do not attach much importance to this. As far as our health goes, we care more for preserving parts and organs of our body — head, heart, extremities, kidneys, liver, and so on. Meanwhile, the appearance of just one unusual cell, a dangerous virus, or bacterium, can pose a real threat to our life. Therefore, the necessity to study individual cells is also dictated by our purely protective reaction, seeking to learn mechanisms of various diseases in order to fight these unwelcome microscopic visitors.

However, it is not just fighting dangerous viruses that stimulates cellular biologists to conduct their research. The very beginning of a new human life, originating in the fusion of two cells — oocyte and sperm — is still full of mysteries. Even though it has been almost two years since (in 1997) the embryologist Ian Wilmut and his colleagues (at Roslin Institute near Edinburgh, Scotland) managed to create a frisky lamb named Dolly from a cell isolated from an adult ewe's mammary gland, their article in *Nature* still reads as science fiction. Indeed, Dolly is a carbon copy, a laboratory counterfeit, so exact that it is in essence its mother's identical twin. The ability to clone adult mammals opens up a myriad of exciting possibilities, from producing replacement organs for transplant patients to cloning champion cows.

To create Dolly, the Wilmut team faced a necessity to manipulate isolated cells without damaging them. Thus, for instance, at one of the crucial stages of cloning, two cells are placed next to each other, and an electric pulse causes them to fuse together like soap bubbles. Although there exist a number of protocols that can bring cells into contact, there is an obscure but highly convenient phenomenon called *dielectrophoresis* that can cause cells to become aligned into long chains of cells called *pearl chains*. This phenomenon is caused when a weak alternating electric field is induced in the aqueous medium containing the cells. This utilizes the same electrodes that carry the fusion-inducing electric pulse. Once cell-to-cell contact is induced, a strong but brief direct current pulse is applied, causing the cells to fuse.

The manipulation of living cells in electric fields is now being developed to produce new hybrid cells; in particular, human hybridoma cells (hybrids between B lymphocytes and myeloma cells). The aim is to produce hybridomas secreting human antibodies to clinically important antigens such as Hepatitis B and other viruses. Human antibodies, unlike the already available animal-derived antibodies (mostly from mice), could be used as therapeutic agents. Mouse antibodies cannot be used in therapy in humans as the body's immune system recognizes these antibodies as nonhuman, leading to severe adverse immunological reactions (including death).

Principles of Dielectrophoresis

The possibility of manipulating the living cells in electric field is explained by their dielectrophoresis. Dielectrophoresis is defined as the lateral motion imparted on the uncharged but polarizable particles as a result of polarization induced by nonuniform electric fields (Pohl, 1978). When a polarizable particle (or cell) is exposed to an electric field, the particle polarizes, giving rise to an induced dipole moment. The value of the dipole depends on the particle volume, its permittivity relative to that of the surrounding medium, and the electric field intensity, E. In the case of a spherical homogeneous particle of radius r the effective dipole moment, m, can be expressed as in Sauer (1985):

$$m = 4\pi\varepsilon_m f({\varepsilon^*}_p, {\varepsilon^*}_m) r^3 E \tag{5.1}$$

where $f({\varepsilon^*}_p, {\varepsilon^*}_m) = ({\varepsilon^*}_p - {\varepsilon^*}_m)/({\varepsilon^*}_p + 2{\varepsilon^*}_m)$ is the so-called Clausius-Mosotti factor, and ${\varepsilon^*}_m$ and ${\varepsilon^*}_p$ are the complex permittivities of the medium and the particle, respectively. A general complex permittivity is given by $\varepsilon^* = \varepsilon - j(\sigma/\omega)$, where ε is the real permittivity, σ is the conductivity, $j^2 = -1$ and ω is the angular frequency. When $\mathrm{Re}\{f({\varepsilon^*}_p, {\varepsilon^*}_m)\} > 0$, the effective moment is aligned with the electric field vector E. Conversely, when $\mathrm{Re}\{f({\varepsilon^*}_p, {\varepsilon^*}_m)\} < 0$, then the effective moment acts against the applied field vector. The dielectrophoretic force F acting on a dipole can be generally expressed as follows:

$$F = \mathrm{Re}\{m\nabla E\} \tag{5.2}$$

Combining Eqs. (5.1) and (5.2), an expression for the dielectrophoretic force acting on a sphere can be written as

$$F = 2\pi r^3 \varepsilon_m \mathrm{Re}\{f({\varepsilon^*}_p, {\varepsilon^*}_m)\nabla E^2\} \tag{5.3}$$

Note that for the case of the sphere the Clausius Mosotti factor is bounded by the limits

$$1 \geq \mathrm{Re}\{f({\varepsilon^*}_p, {\varepsilon^*}_m)\} \geq -1/2$$

As outlined, the Clausius Mosotti factor can be either positive or negative (or zero); so the force on a particle can act to direct a particle either towards or away from a region of high electric field strength. These two conditions are shown in Figure 5.1.

Equation (5.3) is valid for a homogeneous particle. However, we know that living cells are far from homogeneous. Namely, they consist of cellular membrane and cytoplasm. Assume that a spherical living cell consists of a cytoplasm with permittivity ε_p, conductivity σ_p, and radius r enclosed by cellular membrane of permittivity ε_c and thickness δ. As shown by Kaler and Jones (1990), at frequencies higher than 10 kHz the sinusoidal steady-state

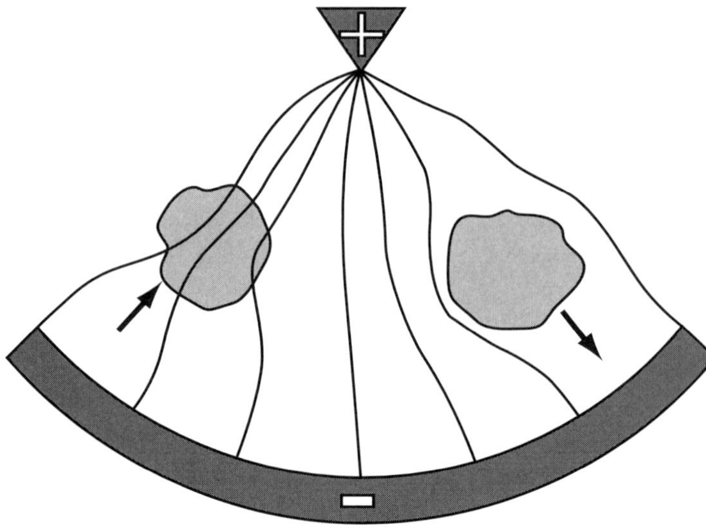

FIGURE 5.1. Two different particles (cells) in a nonuniform electric field. The cell on the left is more polarizable than the surrounding medium and is attracted towards the strong field at the pin electrode; the cell of low polarizability on the right is directed away from the strong field region.

dielectric response of a spherical living cell can be represented by an equivalent homogeneous sphere of radius $r + \delta$ and the complex permittivity $\varepsilon^*{}_{\text{cell}}$, where

$$\varepsilon^*{}_{\text{cell}} = c_m r(j\omega\tau_1 + 1)/[j\omega(\tau_1 + \tau_2) + 1] \tag{5.4}$$

Here, $c_m = \varepsilon_c/\delta$ is the cell membrane capacitance per unit area, and $\tau_1 = \varepsilon_p/\sigma_p$ and $\tau_2 = c_m r/\sigma_p$ are time constants.

As follows from Eq. (5.3), the force acting on a cell in a nonuniform electric field is dependent on its dimensions. Besides, as derived from the definition of the Clausius-Mosotti factor, the force depends on dielectric properties of cytoplasm as well. Thus, mobility of cells in the electric field is a rather complex function of their dimensions and shape as well as of dielectric parameters of cytoplasm and cellular membrane. Thereby, the dependence of the Clausius-Mosotti factor on the electric field frequency allows for a wide range variation of the force acting on a cell.

Now let us return to the problem we posed at the beginning of this chapter: How can we bring the two living cells to be fused so close that they touch? Assume two cells of the same type are close to each other, but out of contact. If the electric field is induced in the solution so that the line joining the cells is parallel to the field strength vector (Figure 5.2), the cells become polarized. As a result, if the cells are located close enough to each other they become attracted to each other due to the interaction of opposite poles of their dipoles. As this theoretical development implies, bringing cells closer can be speeded up if a region of the minimal electric field strength is placed between them, whereas their dipole moment is rendered negative by way of appropriate selection of the field frequency.

Cell ID in an Alternating Electric Field

However, using dielectrophoresis for bringing cells closer before their fusion is but one of its possible applications. At present the most popular field where dielectrophoresis is used appears to be the cell separation. Cell separation has numerous applications in medicine and biotechnology. For example, the isolation of cancer cells from blood and bone marrow is critical

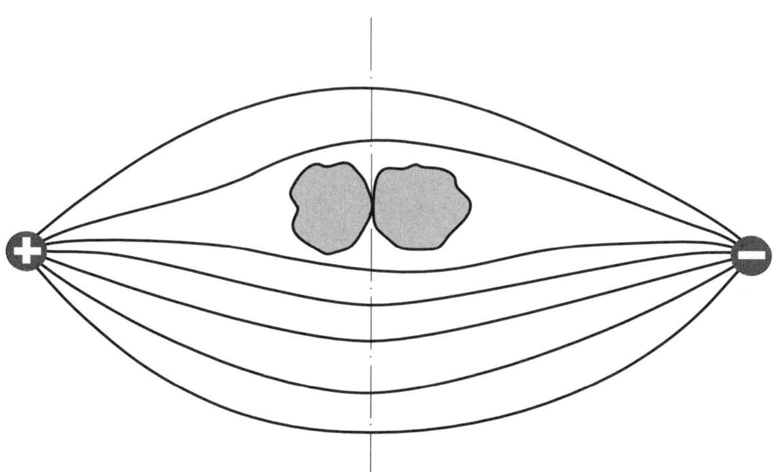

FIGURE 5.2. Bringing cells with negative dipole moments closer in the region of minimal field strength. Mediated by dielectrophoretic forces.

for the early diagnosis of disease and for its monitoring during and after therapy.

As follows from the principles of dielectrophoresis, the electromechanical response of the cell is a complex function of the applied field frequency as well as dielectric parameters of the cell such as the cell membrane dielectric constant and conductance, the conductance and dielectric constant of the cytoplasm, cell size, etc. This implies that, in principle, different cells can be separated from each other by being placed in an alternating electric field, after determining their dielectric parameters. As a rule, the frequency dependence of dielectric parameters of living cells is studied through the use of electrorotation (Arnold and Zimmermann, 1988).

To make cells rotate in alternating electric field without displacing them relative to the optical axis of microscope lens, we use the system of electrodes shown in Figure 5.3, with a 90 degree phase difference between adjacent electrodes. As a result, a rotating electric field is generated, and the resulting rotational torque exerted on a particle is given by:

$$T = m \times E \tag{5.5}$$

Equation (5.5) shows that the torque depends only on the electric field vector and not on the field gradient. The value of the phase difference between the induced dipole m, and the field vector E controls the magnitude of the torque, reaching maximum when the phase difference is 90 degrees, and zero when the phase is zero. Thus a cell in a rotating electric field will rotate asynchronously with the field. It can be shown that the torque depends only on the imaginary component of the dipole moment and so the time-averaged torque on a cell of radius r is:

$$T(\omega) = -4\pi\varepsilon_m r^3 \mathrm{Im}\{f(\varepsilon^*{}_p, \varepsilon^*{}_m)\}E^2 \tag{5.6}$$

For example, if the imaginary component of m is positive, then the torque exerted will be negative and cause the particle to rotate in antifield direction (Figure 5.3(b)).

It follows from Eq. (5.6) that if a cell is placed between electrodes as shown in Figure 5.3, and the angular rate of rotation of cells in the rotating field is measured, the dependence of the angular rate on electric field frequency will coincide with the frequency dependence of the imaginary part

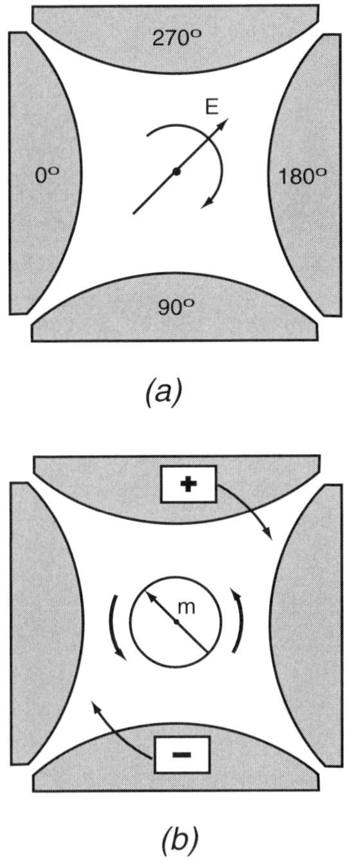

FIGURE 5.3. (a) A rotating field can be generated between four electrodes by applying sinusoidal voltages to them with phases spaced 90° apart. (b) Depending upon the phase angle between the induced dipole moment *m* and *E*, defined by Eq. (5.2), the rotational torque acting on the cell will be either co- or anti-field. In this case, the cell will rotate counter to the clockwise rotating field. Spacing between electrode tips is about several hundred microns.

of the Clausius-Mosotti factor $f(\varepsilon^*_p, \varepsilon^*_m)$. With the frequency dependence for $\mathrm{Im}\{f(\varepsilon^*_p, \varepsilon^*_m)\}$ known, that for $\mathrm{Re}\{f(\varepsilon^*_p, \varepsilon^*_m)\}$ is found fairly easily, as it was done, for instance, in Becker *et al.* (1995) for human breast cancer cells, erythrocytes, and lymphocytes (see Figure 5.4). The frequency dependence of the Clausius-Mosotti factor has proven to be different for different types of cells, and, therefore, it can serve as their ID for conducting the electromechanical manipulations in electric fields.

Cell Separation Using Traveling-Wave Dielectrophoresis

The system of two rows of electrodes given in Figure 5.5 makes it possible to obtain a traveling wave of electric field in the space between the rows. As shown by Huang *et al.* (1993), the time-averaged force $F(\omega)$ acting on a particle in the center of the channel formed between the electrode arrangement pictured on Figure 5.5 can be written as

$$F(\omega) = -\mathrm{Im}\{m(\omega)E\pi/\lambda \tag{5.7}$$

where E is the field strength across the channel and λ is the wavelength of the traveling field of value equal to the distance (measured along the channel) between the closest electrodes of the same phase. Obviously, away from the center of the channel and near the electrode tips, the force acting on the particle becomes more complicated and not describable by Eq. (5.7). The dielectrophoretic force F is balanced by viscous drag (given by Stoke's equation). So, considering the traveling-wave dielectrophoresis in a medium of viscosity η, the velocity u of a cell traveling along the array is represented by:

$$u = -2\pi\varepsilon_m r^2 \mathrm{Im}\{f(\varepsilon^*{}_p, \varepsilon^*{}_m)E^2/(3\lambda\eta) \tag{5.8}$$

The motion of cells along the channel, induced by traveling-wave dielectrophoresis and described by Eq. (5.8), enables separation of living cells from the dead ones. Living cells, with the imaginary part of dipole moment being positive, move to the left; dead ones, with the same being negative, move to the right (see Figure 5.5). Bear in mind here that only those cells that have negative real part of dipole moment will move along the channel between the electrodes since only these cells will strive towards the central axis of the channel, where the region with minimal field intensity is located. Therefore, once the field is on, all the cells possessing positive dipole

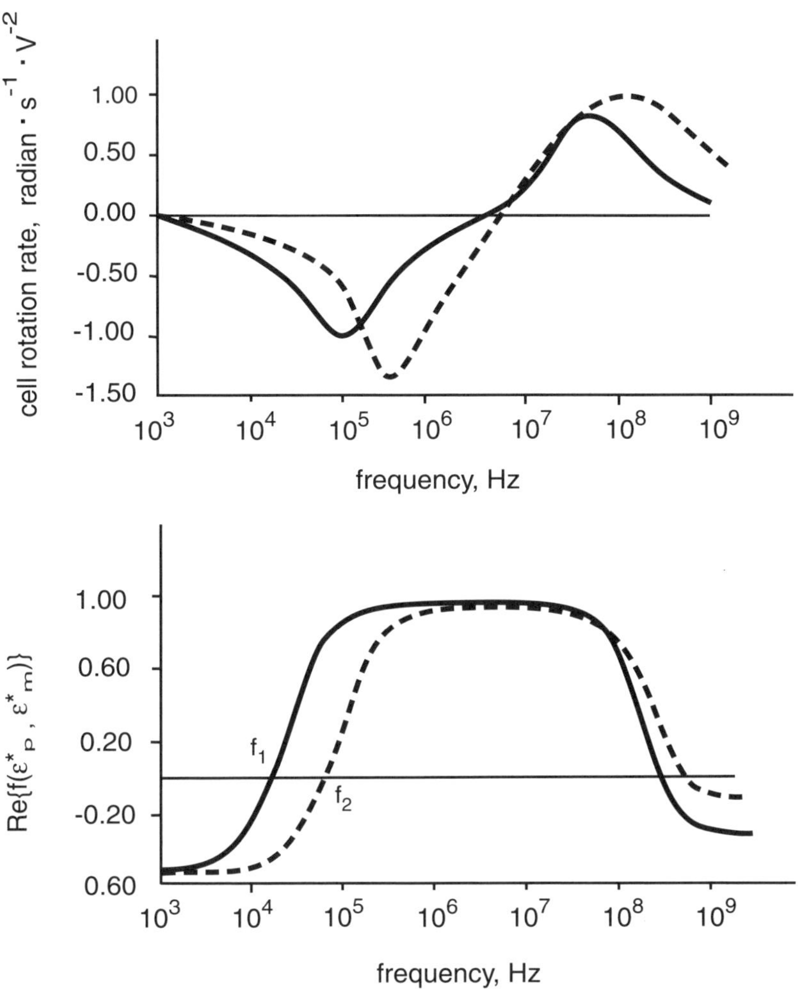

FIGURE 5.4. *Top*: frequency dependence of angular rate for rotation of human breast cancer cells (—) and lymphocytes (- - - -) in isotonic sucrose with conductivity of 56 mS/m. *Bottom*: frequency dependence of Re$\{f(\varepsilon^*_p, \varepsilon^*_m)\}$ calculated from data shown on upper panel; f_1 and f_2 are frequencies at which Re$\{f(\varepsilon^*_p, \varepsilon^*_m)\}$ goes through zero for cancer cells and lymphocytes, respectively. (Modified from Becker *et al.*, 1995. Republished with permission of the Proceedings of the National Academy of Sciences USA, 2101 Constitution Ave., NW, Washington, DC 20418. "Separation of human breast cancer cells from blood by differential dielectric affinity" (Figure 2). F. F. Becker, *et al.* 1995, Vol. 92. Reproduced by permission of the publisher via Copyright Clearance Center, Inc.)

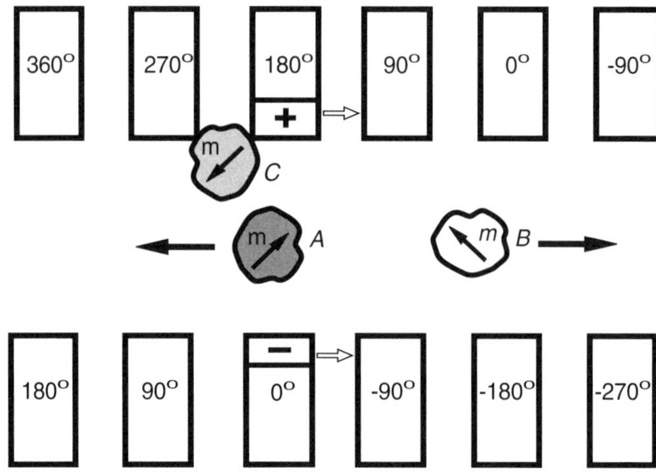

FIGURE 5.5. Cell motion (closed arrows) induced by an electric wave that travels from left to right (open arrows) between electrode system energized by cosine voltage of the indicated phase relationships; *A*—motion expected for a viable cell for field frequency at which Re$\{m(\omega)\} < 0$ and Im$\{m(\omega)\} > 0$ simultaneously; *B*—motion of a nonviable cell with damaged membrane for field frequency at which Re$\{m(\omega)\} < 0$ and Im$\{m(\omega)\} < 0$ simultaneously. *C*—a cell with positive Re$\{m(\omega)\}$ trapped by dielectrophoresis near an electrode tip.

movement at a given frequency will get stuck to the electrodes, where the field strength is, naturally, greater.

This algorithm for cell separation using traveling wave dielectrophoresis makes it possible not only to separate living cells from the dead ones, but also to perform similar manipulations with living cells, differing in their dielectric ID. Thus, for instance, the difference in dielectric ID between human breast cancer cells and lymphocytes enables separating the cells, using an electrode system pictured in Figure 5.5. Indeed, if we apply a field with a frequency in a range between f_1 and f_2, then the cancer cells with a positive real part of dipole moment will go to electrode tips, where they will be trapped near a zone of maximal field strength. Lymphocytes, as shown in Figure 5.5, have a dipole moment with the negative real part at the same frequency, and will move along the channel and be collected at its end.

There could be more sophisticated electrode systems allowing the separation of cells with different IDs. For example, one of the systems

(Figure 5.6), which is a modification of traveling-wave dielectrophoresis, gives an opportunity to separate cell suspension consisting of two cell subpopulations, using two activating frequencies simultaneously. The size of cells to be manipulated dictates the electrode dimensions.

The system pictured in Figure 5.6 is only several millimeters in overall length. The individual electrodes that comprise the active part of the traveling-wave dielectrophoresis array are typically 5 or 10 micrometers in width. The electrode systems are fabricated by using specific combinations of conventional photolithography and/or electron beam lithography.

Electroporation of Cell Membranes

Under normal physiological conditions, a cell membrane is a good barrier for ions and hydrophilic molecules. The membrane specific conductance to Na^+ or K^+ is typically in a range between 10^{-5} and $10^{-3}\,S{\cdot}cm^{-2}$. When an intense transmembrane electric field is applied, the membrane specific conductance increases dramatically, and can reach up to $1\,S{\cdot}cm^{-2}$ in microseconds. The phenomenon can be reversible and has been known since the 1940s as electroporation of cell membranes. For artificial bilayer lipid membranes, the dielectric strength (maximal voltage which can be applied

FIGURE 5.6. An example of an electrode system that uses traveling wave dielectrophoresis to separate (and collect) cell suspension consisting of two cell subpopulations (closed and open circles). Field frequencies F_1 and F_2 correspond to effective motion of empty and closed cells, respectively, at which $Re\{m(\omega)\} < 0$ and $Im\{m(\omega)\} > 0$ simultaneously (see cell *A* on Figure 5.5).

across the membrane divided by its thickness) is determined by lipid composition and is in the range of 300–1,000 kV/cm. Cell membranes can sustain as much as 2,000 kV/cm when microsecond to millisecond electric pulses are used. However, despite extensive information that had already been obtained in early studies, the mechanisms of electroporation of cell membranes remain unclear. It is generally believed that during the electroporation, a rapid structural rearrangement of the membrane occurs, whereby many aqueous pores perforate the membrane.

Electroporation increasingly is used for the introduction of water-soluble molecules into cells, and it has many potential applications in medicine and biotechnology. In the future, for example, the electroporation could be used to facilitate the entry of a drug that suppresses tumor growth into cancer cells. The efficiency of such an action is a thousand times greater that the effect of adding the drug to a suspension of cancer cells. However, a wider application of electroporation is hindered by the absence of a universally accepted model of this phenomenon, although several relevant theoretical studies have already been completed (see Freeman *et al.*, 1994).

6

How Nature Listens

Almost everything that happens on earth produces sound. Sound exists and penetrates everywhere. In contrast to light, sound is able to overcome solid and nontransparent obstacles as well as to skirt them easily. Like light, sound can vary in intensity within a vast range, stimulating relatively simple and highly sensitivity mechanical receptors of the acoustic system of animals. It can provide an animal with the information about the events happening in places that an eye is unable to see. Sound informs a predator that a victim is nearby and gives the latter the last chance to avoid the dangerous meeting. Thus it's no surprise that sound plays a key role in the life of all backboned animals as well as the most active invertebrates—numerous insects.

Let's Recall the Basics of Acoustics

The main parameters of an acoustic wave are its frequency, amplitude (intensity), and the speed of propagation. The frequency of an acoustic wave is fully determined by the properties of the sound source and the speed of its movement relative to the sound receiver (Doppler effect). The amplitude of sound in a given point depends not only on the power of, and the distance to, the source, but also on the properties of a medium. The propagation speed of an acoustic wave is a parameter depending only on the properties of a medium in which the wave expands. Thereby, the pressure in the sound wave fluctuates relative to the value of the average atmospheric pressure, the relative amplitude of these vibrations usually not exceeding 0.5%.

Inside an acoustic wave advancing with the rate c there exists the following ratio between the instant values of pressure P and speed v of the air movement:

$$v = P/\rho c \tag{6.1}$$

where ρ is the air density and the pressure of the sound wave P is interpreted as the difference between the true pressure in the given point and the average atmospheric pressure in the medium.

The product ρc is referred to as *acoustic impedance* of the medium. The higher the acoustic impedance of the medium, the lower the speeds of air movement will cause the same sound pressure in it. Table 6.1 gives the ρc values for air, water, and some biological tissues. Taking the acoustic impedance into account is necessary when analyzing the acoustic wave transfer between different media.

When sound meets the interface of two media, part of the acoustic wave is reflected from the interface while another part transmits into the new medium. We perceive the reflected acoustic wave as an echo. If the vector of the falling acoustic wave speed is perpendicular to the plane of the interface between the media 1 and 2, then the following relationships hold between

Table 6.1. Acoustic impedances of air, water, and some biological tissues.

Materials	Characteristics		
	ρ density $\times$ 10^3 kg/m^3	c, speed of sound $\times$ 10^2 m/sec	ρc, acoustic impedance, N·sec/m^3
Air	0.00129	3.31	430
Water	1	14.8	$1.48 \cdot 10^6$
Muscle	1.04	15.8	$1.64 \cdot 10^6$
Fat	0.92	14.5	$1.33 \cdot 10^6$
Bone	1.9	40.4	$7.68 \cdot 10^6$

the amplitudes of pressures in the incident (P_i), reflected (P_r), and transmitted into the phase 2 (P_t) acoustic wave:

$$P_t = 2\rho_2 c_2 P_i/(\rho_1 c_1 + \rho_2 c_2),$$
$$P_r = (\rho_2 c_2 - \rho_1 c_1)P_i/(\rho_1 c_1 + \rho_2 c_2), \qquad (6.2)$$

where $\rho_1 c_1$ and $\rho_2 c_2$ are the values of acoustic impedance of the media *1* and *2*, respectively.

It follows from expression (6.2) that the pressure inside the reflected and transmitted acoustic waves is fully determined by the values of acoustic impedance of the media. Here, the higher the ratio $\rho_2 c_2/\rho_1 c_1$ is, the higher the amplitude of the reflected wave will be.

Each acoustic wave is a directed flow of mechanical energy. In the cases when the pressure of the acoustic wave changes in time sinusoidally, that is, $P(x, t) = P_0(x)\sin(2\pi ft)$, it can be shown that the amount of energy transferred by the wave per unit time through unit area of the surface perpendicular to c is

$$I = P_0^2/2\rho c \qquad (6.3)$$

Using (6.3) and (6.2), we can obtain the expressions for the energy of the reflected and transmitted acoustic waves:

$$I_r = I_i\{(\rho_1 c_1 - \rho_2 c_2)/(\rho_1 c_1 + \rho_2 c_2)\}^2,$$
$$I_t = 4I_i \rho_1 c_1 \rho_2 c_2/(\rho_1 c_1 + \rho_2 c_2)^2 \qquad (6.4)$$

If an acoustic wave transmits from air to water, then $\rho_1 c_1 = 430\,\mathrm{N \cdot sec/m^3}$ and $\rho_2 c_2 = 1480000\,\mathrm{N \cdot sec/m^3}$. Upon substituting these values of acoustic impedance into (6.4), we obtain $I_r = 0,999\ I_i$ and $I_t = 0,001\ I_i$. Thus, in passing from air into water, 99.9% of the total acoustic energy is reflected from the water surface. The same is evidently true for the acoustic wave passing from water into air.

Finally, the last term in acoustics that we will need to know to evaluate the acoustical system of animals is the average amplitude A of displacements of the medium molecules under the propagation of the acoustic wave. Using (6.1) for sinusoidal acoustic waves we obtain

$$A = P_0/2\pi f \rho c \tag{6.5}$$

where f is the sound frequency.

The Ear in Brief

Figure 6.1 shows a schematic of a section through the head, revealing the main components of the ear: outer, middle, and inner ear. The outer ear, or pinna, acts like a funnel to collect sound and sends it down the ear channel. The outer two thirds of the ear channel is lined with cartilage and contains the sebaceous and wax glands. The ear normally produces enough wax to trap foreign objects. As more wax is produced, the older wax is gently pushed out to the opening of the ear.

Elements of the outer ear serve to direct the energy of acoustic waves to the eardrum, a membrane completely closing the ear channel at its end. The eardrum is a thin layer of tissue about 1 cm (0.4 inches) in diameter. As sound strikes the eardrum, it vibrates and in turn transmits these vibrations to the middle ear.

The middle ear is an air-filled cavity behind the eardrum containing the three smallest bones in the human body (Figure 6.2). These bones, named for

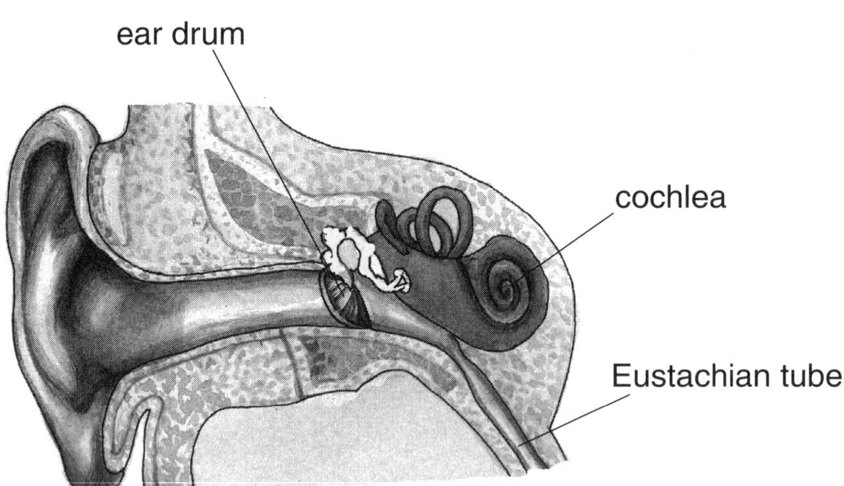

FIGURE 6.1. Schematic section through the head, showing the main components of the ear: outer, middle, and inner ear.

their shapes—the *malleus* (hammer), the *incus* (anvil), and the *stapes* (stirrup)—vibrate along with the eardrum. The combination of the eardrum and the lever action of these bones helps to amplify the sound vibration. The end of the stapes is called the *footplate* and it is connected to the inner ear. As the stapes vibrates, the footplate moves in and out like a piston and transmits the sound vibrations into the inner ear.

The middle ear cavity also contains the opening of the Eustachian tube—a critical structure of the middle ear system. This tube connects the middle ear space with the upper part of the throat. Very small muscles at the base of the Eustachian tube open the tube every time a person swallows. This provides for the exchanges of the air in the middle ear, equalizing the air pressure and refreshing the oxygen supply to the lining of the middle ear. If the Eustachian tube fails to open regularly, bacteria can be drawn up the Eustachian tube and cause an ear infection, called *acute otitis media*.

The *cochlea* is the portion of the inner ear devoted to hearing. It is a spiraling, fluid-filled tunnel embedded in the temporal bone (Figure 6.3). Within the cochlea, fluid-borne mechanical signals are transformed into the

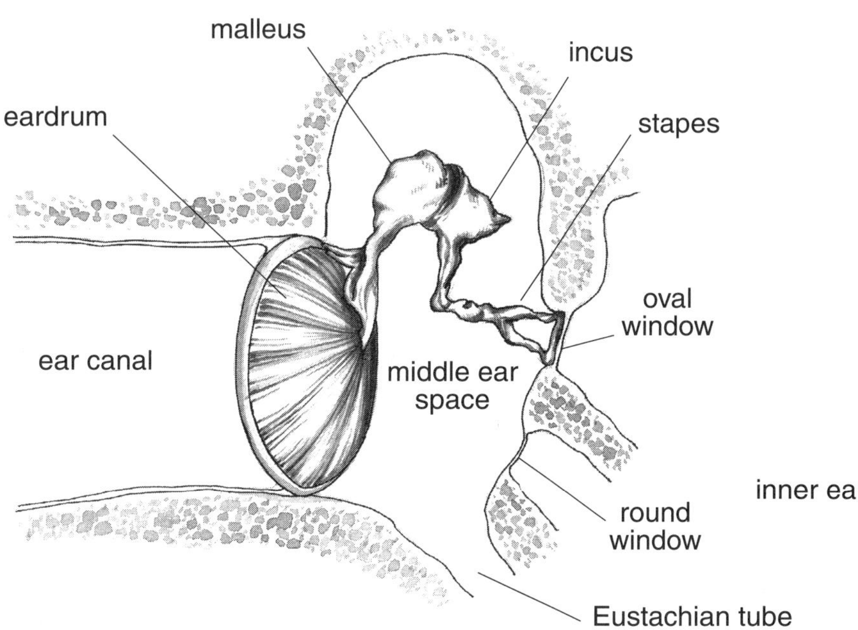

FIGURE 6.2. Enlarged depiction of the human middle ear.

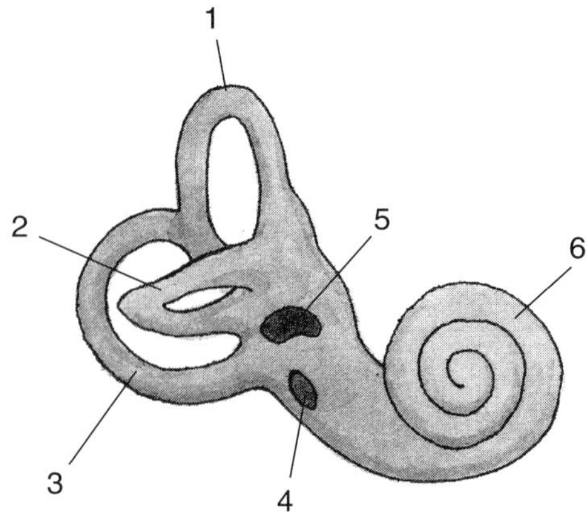

FIGURE 6.3. Inner ear. 1,2,3, superior, lateral, and posterior semicircular channels, respectively; 4, round window; 5, oval window; 6, cochlea.

neural code delivered by the auditory nerve to the brain. When air pressure in front of the eardrum increases, the eardrum is pushed inward, moving the three small bones of the middle ear. The footplate of the stapes covers the oval window of the cochlea. The movement of the stapes initiates a pressure wave in the cochlear fluid. Figure 6.1 shows the location of the cochlea in relation to the parts of the middle ear.

Acoustic signals that enter the fluid-filled cochlear chambers propagate in a disperse manner along the cochlear partition. The partition, which spans the length and width of the cochlea, consists of the basilar membrane, tectorial membrane, and organ of Corti. The organ of Corti is a collection of cells, including the sensory hair cells, that sit on the basilar membrane. Acoustic vibrations that propagate inside the middle ear lead to the appearance of a running wave over the basilar membrane and deform the sensory hair cells, whose excitations result in our acoustic sense.

The Middle Ear

The middle ear is thought to aid the transmission of sound energy from the ear channel to the cochlea by partially matching the impedance of the air in the channel to the impedance of the fluids in the cochlea; the mechanisms involved in this process are not well understood and are only briefly discussed in the next section. In an effort to better quantify the function of the middle ear, different models have been proposed by numerous investigators (see for example, Ladak and Funnell, 1993).

The transmission of sound from one medium to another is reduced if the acoustic impedance of the two media are different; full transmission is possible only if the impedance is matched (see Eq. (6.4)). Since the impedance of air is much lower than that of the cochlear fluids, direct stimulation of the cochlea by sound would not be very effective in transmitting energy to the sensory receptors within it. The middle ear improves sound transmission by partially matching the impedance of these two media. Such matching results in an increase in the pressure at the stapedial footplate relative to the pressure at the eardrum.

The eardrum of a human being has an area of about $0.7\,cm^2$. Auditory ossicles connect the eardrum to the membrane of the oval window, whose area is only $0.03\,cm^2$. The ratio of the area of the eardrum to that of the footplate is thought to be the main mechanism involved in this pressure transformation process. Sound pressure acts over the large surface of the eardrum, and the resultant force is applied to the smaller area of the footplate. Therefore, the pressure at the footplate will be larger than that at the eardrum. The ratio of the eardrum area to the area of the footplate is about 15 in humans.

The lever consisting of middle ear bones and the curvature of the eardrum can also contribute to the pressure transformation process. The malleus and incus, rotating as a single rigid body, form a lever that serves to amplify the force acting on the eardrum before it is applied to the stapes. Also, as was originally proposed by H. L. von Helmholtz (1869) and presently developed by Funnell (1996), the curvature of the eardrum amplifies the force that acts on its surface before it is applied to the malleus. It is not clear how much each of the mechanisms contributes to the total value of the middle-ear transformer ratio.

Thus, the middle ear operates as a pressure transformer, increasing the acoustic pressure on the membrane of the oval window approximately 40 times compared to the pressure on the eardrum. In this process the amplitude of the eardrum vibrations exceeds that of the oval window by approximately twice. It follows from Eq. (6.1) that the acoustic impedance is equal to the ratio of the pressure amplitude to the amplitude of the displacements of the medium particles in a given point during the acoustic wave advance. From what we have said concerning the functioning of the middle ear, it follows that the acoustic impedance of the ear at the level of the eardrum is 80 times lower than that at the level of the oval window, and has a value close to the acoustic impedance of air. The proximity of the values of acoustic impedance of the middle ear and air leads to a significant decrease of the amplitude of the acoustic wave reflected from the eardrum. Thus the middle ear matches the acoustic impedance of the middle ear and that of air.

The ear of terrestrial animals is tuned to be receptive of acoustic waves in air, and thus operates poorly under water. This is due to the fact that the acoustic impedance of water is approximately 1,000 times greater than that of the middle ear cavity filled with air. Therefore almost all sound is reflected from the eardrum when we are under water.

The imperceptibility of the underwater sounds for the human ear gave our ancestors reasons to think that the undersea world is a world of silence. The fact that fish can interact with the help of sounds became known only in the 1940s, when the development of submarines led to the creation of special hydroacoustic systems.

However, it was Leonardo da Vinci who proposed to listen to the undersea sounds by applying the ear to an oar put in water. The acoustic impedance of wet wood is close to that of water, and the narrowing of the oar from a blade to a handle makes the oar an acoustic transformer analogous to the middle ear. As a result, the undersea acoustic waves advance along the oar, with an insignificant reflection, to the cranium bone close to the ear. Vibrations of this bone cause analogous vibrations of the liquid in the inner ear and, thus, acoustic senses. Fishermen, who listen to the undersea sounds in such a manner, know that fish are very talkative.

The ears of whales and dolphins are well-tuned to the undersea sounds. Unlike terrestrial mammals, the connection of acoustic bones and their dimensions in sea mammals are such that the amplitude of the vibrations for the membrane of the oval window significantly exceeds that of the eardrum. As a result the acoustic impedance of the ear at the level of the eardrum increases, becoming closer to the impedance of water. Thus the middle ear of whales and dolphins, as that of terrestrial anaimals, has the value of the acoustic impedance close to that of the acoustic medium, which enables a larger acoustic energy to be transferred into the inner ear.

Cochlear Amplifier

The optimal structure of the middle ear and the high sensitivity of the hair cells in the inner ear enable most animals to sense the low-amplitude acoustic vibrations beyond the sensitivity limits of modern acoustic systems. The minimal sound intensity that the human ear is able to distinguish is $10^{-12}\,\mathrm{W/m^2}$.

To get an idea about the sensitivity of our ear, using Eq. (6.3) and (6.5), let us find the corresponding amplitude ($A_{\min}$) of sinusoidal movements of air molecules in propagating acoustic wave. Substituting $I = 10^{-12}\,\mathrm{W/m^2}$, $\rho = 1.3\,\mathrm{kg/m^3}$, $c = 330\,\mathrm{m/sec}$, $f = 3\,\mathrm{kHz}$ yields a value of $A_{\min}$ close to $10^{-11}\,\mathrm{m}$. For comparison, let us note that the diameter of a hydrogen atom approximately equals $10^{-10}\,\mathrm{m}$.

Thus the minimum average amplitude of sinusoidal movements of air molecules in an acoustic wave perceivable by humans is only one tenth of the hydrogen atom diameter. We would probably gain no advantage if our ear became several times more sensitive since, with such sensitivity, this organ would react to random Brownian movements of air molecules. In this case we would hear the meaningful sounds against the background of continuous hissing ("white noise"), which bears no useful information. To evaluate the force (intensity) of sound L, besides (6.3), the following value is often used:

$$L = 20\,\lg(P/P_{\min})$$

where P is the amplitude of acoustic pressure of a given wave, and $P_{\min}$ is the amplitude of acoustic pressure corresponding to the acoustic threshold of the human hearing. Table 6.2 gives the values of L corresponding to the different values of P.

One of the reasons for the phenomenal sensitivity of our ear is the structure of the cochlea — the main element of the inner ear. The manuscript by Dallos *et al.* (1996) presents a detailed overview of the mammalian cochlea — from its anatomy and physiology to its biophysics and biochemistry.

The cochlea consists of a single bony tube that spirals around a middle core containing the auditory nerve. Similar to a snail, the spaces are larger in the basal turn than they are in the apex. At each turn, the bony tube is divided into three separate compartments by the membranous tissues. A

Table 6.2. Sound power and amplitude of sound pressure.

Sound pressure amplitude, N/m^2	Sound intensity, dB	Real life manifestations
2000	160	Mechanical damages of eardrum
200	140	Pain threshold of ear
2	100	Working factory, car
0.02	60	Office noise, conversation
0.0002	20	A very quiet room
0.00002	0	Audibility threshold

cross section through one cochlear turn shows how each turn of the cochlea is divided into three chambers, called *scalae* (see Figure 6.4). Scala tympani and scala vestibuli contain perilymph. Scala media contains the endolymph, which is bounded on three sides by stria vascularis, Reissner's membrane, and the organ of Corti. The organ of Corti contains the sensory hair cells that transduce mechanical stimuli (sound) into electrical signals.

Motion of stapes sets an oval window (see Figure 6.2) and then fluid in the cochlea into motion. Vibrations produce a traveling wave in the basilar membrane. As fluid is noncompressible, there must be another mobile wall to permit vibration. A round window serves this function. Both the cross-sectional area of the scala vestibuli and the mechanical properties of the basilar membrane change along the direction of the propagation of the wave; therefore, on the way from the oval window to the cochlea apex, the amplitude of vibrations of the membrane changes, too. The rigidity of the basilar membrane decreases from the stapes to the cochlea apex. Thus the propagation speed of the wave gradually falls while approaching the cochlea apex, and the wavelength decreases. For the same reason, the amplitude of the wave that propagates from the stapes to the cochlea apex initially increases, becoming significantly larger than when in the proximity of the stapes. Subsequently, upon achieving the maximum value, the amplitude of vibrations of the basilar membrane decreases due to energy dissipation in the liquids of the inner ear. In this case the position of the maximum of vibrations of the membrane depends on its frequency. At higher frequencies, it is closer to the stapes, and at lower ones, to the cochlea apex (see Figure 6.5).

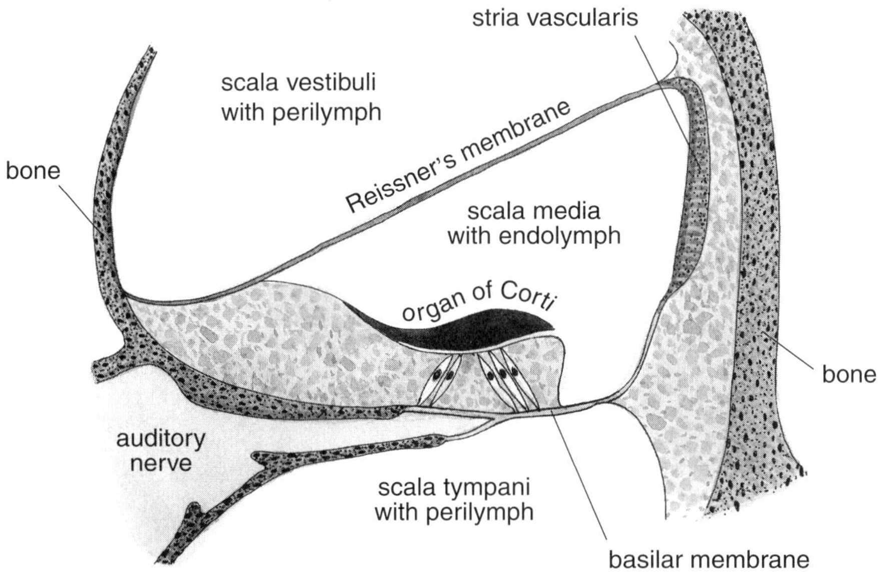

FIGURE 6.4. Cross-section through one cochlear turn.

Therefore, the maximum vibrations of the basilar membrane for each frequency of the acoustic signal are detected at a *specific* distance from the stapes, different for different frequencies. Each fiber of the acoustic nerve originates from the hair cells located in a different part of the cochlea. Thus the maximum activity of this fiber emerges as a response to the action of the frequency that is characteristic of this particular part of the cochlea. In other words, cochlear nerve fibers that innervate hair cells at different points along the length of the basilar membrane (from oval window to cochlea apex) are tuned to different frequencies of sound. This type of nerve coding is named *place coding*.

Most of these ideas about the hearing mechanism resulted from the works of Georg von Békésy, who received his Nobel Prize in Physiology or Medicine in 1961. Georg von Békésy viewed the role of the human cochlea as purely passive and suggested that final tuning of the response that is necessary to achieve pitch and intensity discrimination occurs in the central portion of the auditory pathway. However, even after the works of von Békésy, many problems are still unsolved. The main problem is that the high

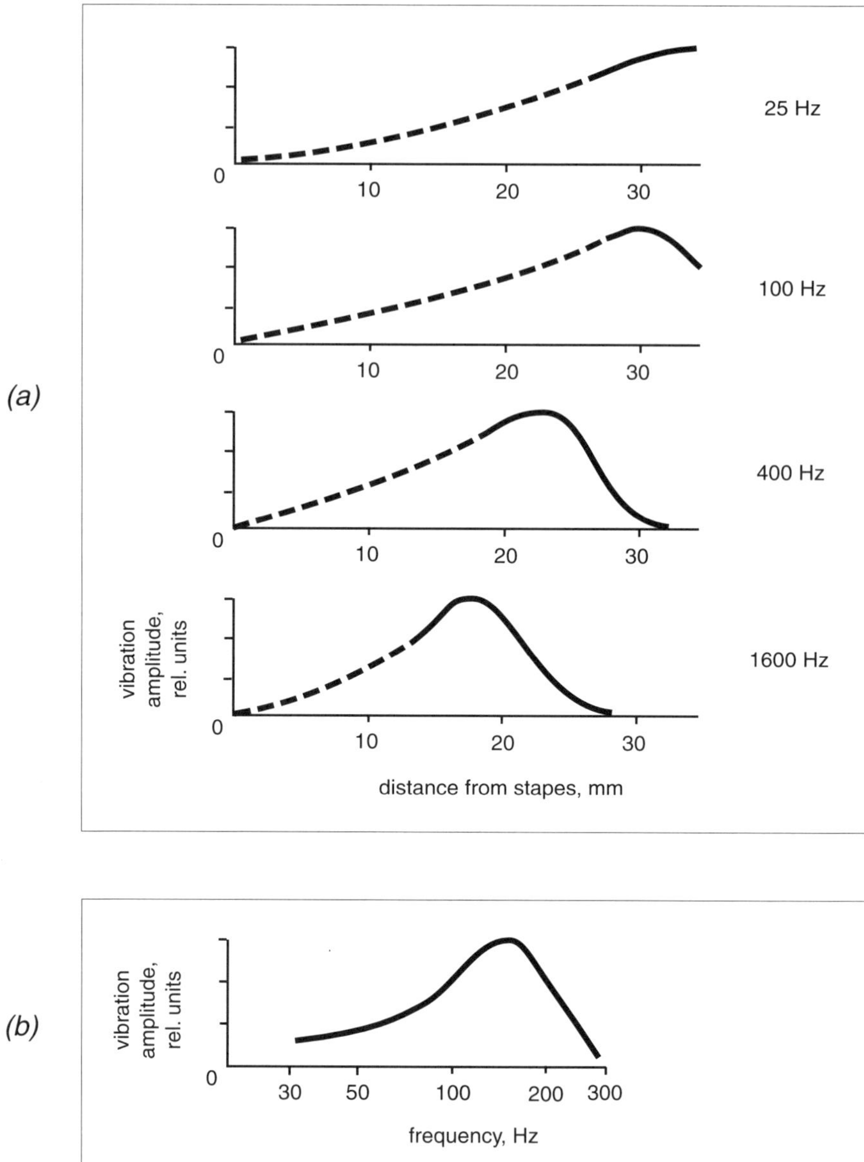

FIGURE 6.5. Amplitude of vibrations produced by a traveling wave in the basilar membrane at various sound frequencies (a), and frequency dependence of the vibrations on a distance of 30 mm from stapes (b). (Schematic depiction.)

degree of acoustic resolution achieved by the hearing can't be explained solely by passive mechanical properties of the basilar membrane.

Measurements from single auditory nerve fibers, going from different regions of basilar membrane, provide information about cochlear tuning curves. As the measurements show, each point on the basilar membrane is tuned to a different frequency, with a spatial gradient of about 0.2 octaves/mm for humans (the human basilar membrane is about 35 mm long). Roughly speaking, the cochlea acts like a bank of filters. The filtering allows for the separation of various frequency components of the signal with a good signal-to-noise ratio. Typical tuning curves in a human have bandwidths, which are commonly characterized as being about 1/3 octave (compared with a bandwidth of about two octaves for basilar membrane tuning in Figure 6.5). How does the cochlea deliver to each nerve fiber only a specific, narrow range of frequencies?

One of the generally accepted hypotheses associates the fine frequency tuning of our ear with the outer hair cells located in the organ of Corti (Figure 6.6). The organ of Corti is a collection of cells, including the sensory hair cells, that sit on the basilar membrane. Along the upper surface of the organ of Corti, hair bundles protrude from the tops (or apexes) of the hair cells. Each bundle is composed of two to four rows of hair-like structures called stereocilia. Connected to the bottom (or base) of each hair cell are

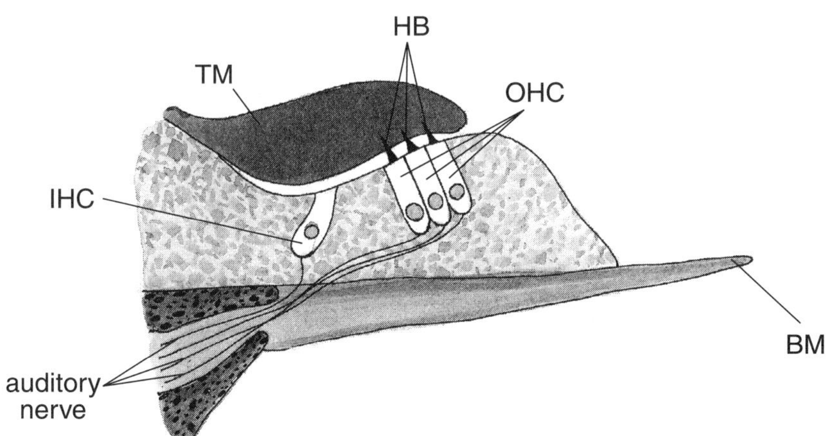

FIGURE 6.6. Organ of Corti. TM (tectorial membrane), HB (hair bundles), OHC (outer hair cells), IHC (inner hair cell), BM (basilar membrane).

nerve fibers from the auditory nerve. There are two types of hair cells in the cochlea. The inner hair cells are primarily innervated (that is, connected to the auditory nerve) by afferent fibers, which deliver neural signals to the brain. The outer hair cells, on the other hand, are innervated primarily by efferent nerve fibers, which receive neural signals from the brain.

The human cochlea is believed to contain approximately 4000 inner and 12,000 outer hair cells, with four cells radially abreast and spaced every 10 microns along the basilar membrane's length. The tectorial membrane lies on top of the organ of Corti with a thin fluid space (4 to 6 microns) separating these two surfaces, which shear as the basilar membrane moves up and down.

Hair cells are primarily mechano-electric transducers that convert displacement of the hair bundle into a change in the receptor current flowing through the cell. This is done by mechanical gating of ion channels that must be located in the hair bundle, probably near the top of each stereocilium.

Outer hair cells have a special function within the cochlea. They have stereocilia at the top of the cell, and when the stereocilia are bent in response to a sound wave, an electromotive response occurs. Displacement of the stereocilia towards the tallest produces a cell depolarization as ionic current flows in through the mechanoelectric transducer channels. The outer hair cell responds to the depolarization by decreasing length. Outer hair cells also act as electromechanical transducers by converting voltage across their cell membrane into length changes. So, with every sound wave, the cell shortens and then elongates. This pushes against the tectoral membrane, selectively amplifying the vibration of the basilar membrane. This capability of outer hair cells allows us to tune finely to different frequencies and hear very quiet sounds. This hypothesis about the crucial role of outer hair cells in active tuning of the cochlea requires additional experimental and model studies.

Ultrastructural analysis of cell lateral plasma membrane of outer hair cells indicates that electromotility may be based upon conformational changes of a dense array of integral membrane proteins whose molecular identity is not yet known. Changes in membrane potential may act as the driving force for the rearrangement of the proteins in the plane of the membrane. Membrane surface forces are converted into cell axial forces by interaction of the motor proteins with the cytoskeleton. A membrane motor model for the fast motility of the outer hair cell proposed by Iwasa (1994) predicts that the force produced under isometric condition is about 0.1 nN/mV, in agreement with values estimated from in vivo conditions.

However, it is not yet completely understood how outer hair cell length changes might translate into an active biomechanical mechanism to enhance sensitivity and tuning. Therefore, if you wish to be involved in this study, we refer you to papers by Hubbard (1993), Ashmore *et al.* (1995), Nobili and Mammano (1996), Ulfendahl (1997), and Brownell and Popel (1998).

Otoacoustic Emissions, or Ear Sounds

Did you know that your ear not only hears but also produces sound? These sounds are called *otoacoustic emissions*. They were accidentally discovered in 1978 by David Kemp, who was trying to build an improved tiny microphone to measure how much sound bounces off an eardrum. Using a signal averaging technique, he identified a new auditory phenomenon in the acoustic impulse response of the human ear. Kemp found that when a brief sound, such as a click, is presented to the ear, a very weak sound similar in form is emitted by the ear after a slight delay (about of 5 ms). The slowly decaying response component was present in all normal ears tested, but was not present in ears with cochlear deafness. Therefore, the otoacoustic emissions are thought to be sounds produced by healthy ears in response to acoustic stimulation. Although Kemp's observations were greeted with skepticism, the otoacoustic emissions have since been reliably confirmed.

As we mentioned, the contractile activity of the outer hair cells, by selectively amplifying the vibration of the basilar membrane, can help the ear to filter sound. Therefore, it was proposed that the otoacoustic emissions could be byproducts of the activity of the outer hair cells in the cochlea and generated by their motile elements. In the past few years, audiology has seen a remarkable advance in objective audiometry techniques due to the advent of otoacoustic emission measurement. This easy-to-use, noninvasive, and objective diagnostic tool delivers reliable judgments of inner ear status by applying low sound stimuli to the ear and recording the resulting emission, or "Kemp echo," from the ear. As a result, otoacoustic emissions testing became the hottest topic in the audiological industry.

All otoacoustic emissions are divided into two broad categories: spontaneous, occurring without the presentation of an activating stimulus;

and evoked, occurring as a direct result of an activating stimulus. The evoked otoacoustic emissions are measured by presenting a series of acoustic stimuli to the ear through a probe that is inserted in the outer third of the ear channel. The probe contains a loudspeaker that generates clicks (or tones) and a microphone that measures the resulting otoacoustic emissions that are produced in the cochlea and that subsequently reflected back through the middle ear into the outer ear channel. The resulting sound that is picked up by the microphone is digitized and processed by specially designed hardware and software. The very low-level otoacoustic emissions are differentiated by the software from both the background noise and the contamination of the evoking stimuli.

Soon after his report of emissions evoked by clicks, Kemp described his discovery of the acoustic emission of intermodulation distortion by the ear. To produce such emission, two sine-wave tones close in frequency ($f1$ and $f2$) are presented to the subject's ear at the same time. As was shown later, the outer hair cells stimulated by $f1$ and $f2$ frequencies stimulate a third set of outer hair cells. The oscillatory contractions of this third set of outer hair cells create a separate tone, known as the distortion product. The corresponding sound then emitted by the ear has been called distortion product otoacoustic emissions.

The strongest of these distortion products is typically at the frequency $2f1 - f2$. For example, if the tones used are at 1000 Hz ($f1$) and 1200 Hz ($f2$), the strongest distortion product will be at 800 Hz. The relative intensity of the distortion product, compared to the two sine waves, can be used to assess the function of the ear in that part of the frequency range.

The otoacoustic emissions by the cochlea and the re-emission of this energy as sound from the ear serves no important physiological purpose that can be determined. Their clinical significance is that they are evidence of a vital sensory process arising in the cochlea. Otoacoustic emissions occur only in a normal cochlea with normal hearing. If there is damage to the outer hair cells that produces a mild hearing loss, then otoacoustic emissions are not evoked. In a typical hearing test, tones of different intensity are played to a person through headphones. The person is asked to indicate that a tone has been heard by raising his or her hand, for example. Obviously, such a test cannot be administered to a baby. However, it is important to know as early as possible if a child has a significant hearing impairment so that steps can be taken to help the child. Therefore, many states are now requiring all infants to be screened for hearing loss. Since the method discovered by Kemp does not require the subject to give a hand signal or to speak, it can be used as a general test for hearing disorders in infants. Such tests, in fact, can be applied to an infant on the day it is born.

Although it is not yet completely understood how the outer hair cells might produce the otoacoustic emissions, in the paper by Talmadge *et al.* (1998), some models that reproduce evoked emissions, distortion-product emissions, and spontaneous emissions were compared, which makes this reference useful.

What Is a Cochlear Implant?

It is estimated that more than two million Americans, and millions more worldwide, are profoundly deaf and may be unable to participate in normal conversation, to hear warning signals (smoke alarms, sirens, etc.), or to use the telephone. Approximately two percent of them are absolutely deaf. How can we help them?

The reason of the absolute surdity, as a rule, is in the injury of the inner ear (cochlea), although the acoustic nerve that connects the ear to the brain often remains vital. Thus, the main strategy to help absolutely deaf people is the electric stimulation of the acoustic nerve.

The acoustic nerve of a human being includes more than 30,000 nerve fibers that transfer information on the vibrational spectrum of the eardrum to the brain. In this process the sequence of nerve impulses in each fiber indicates the presence of a particular harmonic in this spectrum and the phase of its harmonic relative to the others.

A cochlear implant is an electronic device designed to provide sound information for adults and children who have a sensorineural hearing loss in both ears and obtain limited benefit from appropriate binaural hearing aids. The first research on cochlear implants was conducted in France over thirty years ago. Since then, cochlear implant technology has evolved from a device with a single electrode (or channel) to systems that transmit more sound information through multiple electrodes (or channels). Those who used the cochlear implants confirm that the implants restored the hearing to a practically full extent, except in those who were born deaf. By now more than 15,000 people worldwide have received cochlear implants.

Usually, 22 electrodes are inserted inside the inner ear of the patient. They are positioned on the surface of the cochlea in such a way to offer the possibility of separately stimulating nerve fibers that originate from different parts of the basilar membrane up to the place of their merging into the acoustic nerve. As we explained previously, when the spectrum of the acoustic signal contains high frequencies, the highest amplitudes are exhibited by the vibrations of the part of the membrane close to the stapes, and the lower frequencies result in vibrations of its remote inner part. Therefore, for a normal human, the nerves originating from the outer edge of the membrane are stimulated at high sound frequencies, and the ones originating from the opposite edge at lower frequencies.

Common cochlear implant devices necessarily include a speech processor with the dimensions of a pocket calculator. In the simplest cases the speech processor decomposes the signal from a tiny microphone into 22 harmonics and sends the stimuli to the 22 electrodes, each corresponding to the amplitude of the given harmonic in the acoustic spectrum. As a result, the brain accepts the signals of the rough but spectral analysis of the surrounding acoustic field. As a rule, the speech processor enables us to use several strategies, and the patient is able to choose the optimal program of the stimulation. These speech processing strategies can differ in the parameters of the filters used and in other technologies, which are the intellectual property of the producers. In either case, it takes about half a year for the patient to find the optimal speech processing strategy.

It is evident that to stimulate each fiber (or a group of fibers) separately in this way entails significant technical and surgical difficulties. Therefore, sometimes the acoustic nerve is stimulated as a whole with the help of several algorithms. One of the most widespread methods of the acoustic nerve stimulation is preceded by a significant simplification of acoustic vibrations surrounding a deaf person. The sound is transformed by a microphone into electric oscillations, and a special device determines their average frequency. This becomes the frequency at which the generator stimulates the acoustic nerve by the electric impulses of the standard amplitude.

The question legitimately arises, whether such a primitive encoding of the acoustic information cures the deaf person. Certainly, no. However, along with the capability of most deaf people to understand a companion by watching the movements of the lips, the electric stimulation of the acoustic nerve offers numerous improvements. With its help, a deaf person can distinguish an interrogative tone by the higher average frequency of the stimuli. Besides, the stimulation enables the patient to determine more distinctly the beginning and end of the words and phrases of the speaker. In general, this enables a deaf person to increase the rate of lip-reading by 60–70%.

It is obvious that the optimal speech processing strategy has to be rendered insensitive to the noise that bears no useful information, as well as to have subroutines corresponding to various situations (a walker in the street, a driver or passenger in a car, a lecturer or audience, a TV watcher, etc.). The search for the optimal speech processing strategy is one of the possible fields of cooperation between physicists and biologists.

Sound Localization

It is important for many animals not only to hear sound but to determine the direction from which it comes. One of the pioneers in spatial hearing research was J. W. Rayleigh. About 100 years ago, Rayleigh developed his so-called *duplex theory*, stating that there are two principal ways to determine the direction to the source of sound, both requiring two sound receivers (ears). In the first method, the measured quantity is the interaural time difference (ITD) between the arrivals of the same component of the acoustic wave. Let us assume that, from a distant source, a sound wave strikes a spherical head of radius a from a direction specified by the azimuth angle θ (Figure 6.7). In this case, we obtain the following relationship between the parameters

$$ITD = (\theta + Sin\theta)a/c \tag{6.6}$$

where c is the speed of sound.

Obviously, the higher the ITD, the greater the angle between the direction to the source of sound and the average line of the head. The average distance between the human ears is about 0.17 m. Therefore, the maximum value of the delay in propagation of the acoustic wave perpendicular to the midde plane of the head is equal to $0.17/330 = 5 \cdot 10^{-4}$ sec. In fact, the ability of our ears and brain to cooperatively determine the ITD, in less than 10^{-4} sec,

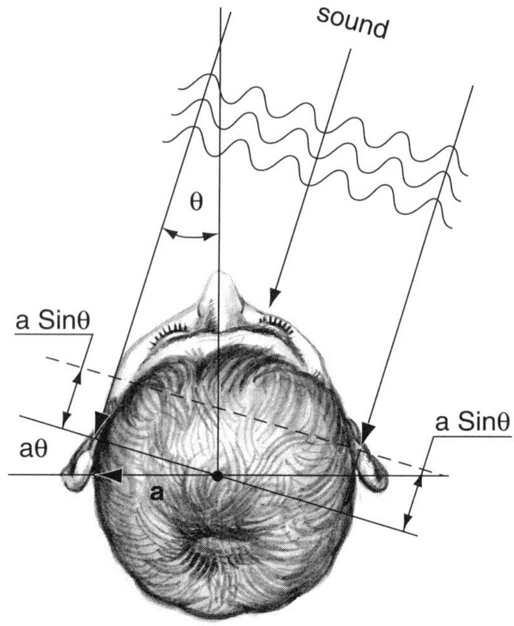

FIGURE 6.7. Calculation of interaural time difference for the model of a human head.

is truly surprising. Well-trained people, for example, can determine the deviation of the source of sound from the midde plane by an angle of less than 1–2 degrees, thus identifying the time delay of about 10 μsec. Obviously, this way of determining the direction to the source of sound is more useful for large animals, whose ears are father from each other.

The second way is to measure the interaural intensity difference. For the acoustic vibrations of the wavelength lower than 15 cm (i.e., frequency higher than 2 kHz) the human head presents a significant obstacle. An acoustic shadow appears behind the head, where the sound intensity is lower than on the opposite side. If the source of sound is shifted relative to the midde plane of the head, then one ear will be in the sound shadow and the other outside the shadow. As a result, the intensities of acoustic waves that reach both ears will be different.

Experience taught us to use this difference in the intensity of sound to determine the direction to the source of sound with an accuracy of 1–2 degrees. Subconsciously, a human uses both methods to localize the source of sound. The 10% difference in the sound intensities corresponds to the 18 μsec delay in the acoustic wave getting to the farther ear. Therefore, if we found a way to make the ear outside the acoustic shadow 10% less sensitive, then the testee systematically would make a 3–4 degree error in determining the direction of the source of the sound.

Note that for small animals (e.g., mice) only the second way of localizing the source of sound is available since the first one would require that they be capable of measuring time intervals with the accuracy of 1 μsec, which is impossible for the nervous system. The unusual sensitivity of small animals to high-frequency sounds creates necessary conditions for the use of the second method. In fact, compared to a human, whose ear is insensitive to frequencies higher than 20 kHz, mice can distinguish sounds at frequencies of up to 70 kHz.

When humans find themselves underwater, close to a moving motor boat, they experience an unusual sensation. Regardless of where the boat is located, it seems that it is somewhere nearby since the audibility of its motor doesn't change in the water when turning the head. Without special devices, a human with the head underwater is unable to determine the direction of the sound's source. This can be explained by the fact that the bones of the cranium don't create the sound shadow in the water since their acoustic impedance is close to that of water. That is why acoustic waves in water reach both ears in the same way, thus depriving a human of the way to determine his/her acoustic orientation by measuring the interaural intensity difference.

Acoustic system of whales and dolphins is well suited to undersea orientation with the use of sound. Air cavities in their cranium produce sound shadows (due to the huge difference in acoustic impedance of air and water). The shadows spatially divide both ears. This enables the animals to determine the direction of the sound's source underwater with the help of these methods.

It could appear that insects must meet insurmountable difficulties in determining the direction to the source of sound. The thickness of a grasshopper's body is less than 1 cm, and the wavelength of the sound it produces is about 8 cm (frequency 4–5 kHz). Therefore the grasshopper itself won't create a sound shadow for the chirping of a neighbor, and even the presence of two acoustic organs located at different parts of the body gives it no opportunity to determine where the neighbor chirps. Nevertheless grasshoppers excellently find each other using acoustic signals only. It seems that what enables them to do this is an unusual structure of their acoustic organs, though not all the mechanisms with which the insects hear are clarified yet (Yack and Fullard, 1993).

Grasshoppers and other insects have their acoustic organs located in the front limbs, somewhere below the knees. Insect ears represent modified chordotonal organs that have a thin cuticular membrane, *tympanum*, that acts like an eardrum (Figure 6.8). Thus, for example, the hearing organs were found in noctuid moths, which use them to avoid predators.

Unlike mammals, an acoustic wave can affect the *tympanum* of a grasshopper from both sides since its inner surface is also in contact with the atmosphere through a narrow hole, a *spiracle* (Figure 6.8). In this case the deviation of the eardrum will be directly proportionate to the pressure difference at both sides of the eardrum. Since a grasshopper has a small acoustic organ compared to the wavelength of sound, the resulting force

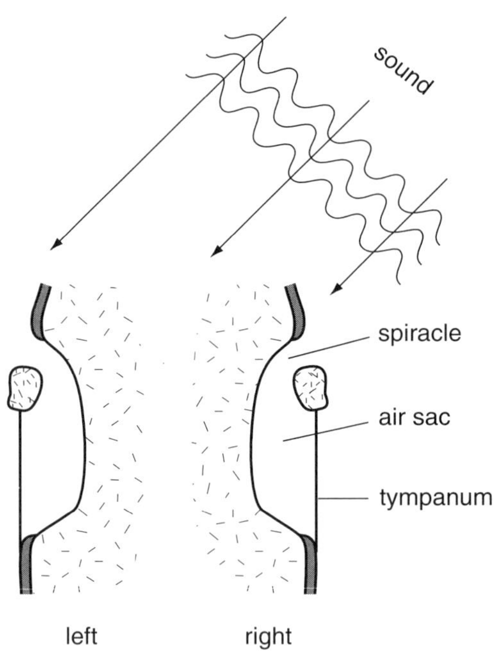

FIGURE 6.8. Section through left and right hearing organs of an insect.

acting on the tympanum will be proportional to pressure gradient in the advancing acoustic wave.

Paired acoustic organs sensitive to pressure gradient enable the grasshopper to determine the direction of the sound's source. When the direction of sound propagation is perpendicular to the plane of the tympanum, the amplitude of the tympanum vibrations is maximum for both organs; however, their vibrations are shifted by 180 degrees in phase. When the direction of the acoustic wave is parallel to the plane of the membranes, then the vibrations of the membranes will, obviously, be absent. Thus, by comparing the amplitude and phase of the membranes of both acoustic organs, the grasshopper determines the position of the source of the sound.

A similar principle of localizing the sound's source can be found in some other small animals. If we consider a frog, we see that an acoustic wave passes through the eardrum of one of its ears, then through the syrinx, acting on the inner surface of the other ear's eardrum. Therefore the acoustic organ of a frog is also sensitive to pressure gradient. Humans have a very narrow syrinx, and as a rule it is filled with liquid. So we can't use such a mechanism to localize the source of sound.

If you want to study the sound localization problem in depth, the book written by Blauert (1997) can be a useful reference on the psychophysics of three-dimensional hearing.

Echolocation

The most vivid examples of the use of sound in animate nature are human speech and echolocation. Echolocation is most developed in bats, who are active at night. For them, echolocation replaces vision. It is interesting that the principles of echolocation had been used in radar and sonar before they were discovered in animals. Yet the skill with which bats derive information, proceeding from the echo of the transmitted signals, is truly fantastic. Analyzing the reflected signals, not only are they capable of locating surrounding objects, but they can also determine the distance to them, their azimuth and dimensions, as well as the type of their surface. The majority of bats are entomophagous. Echolocation helps them hunt mosquitoes, which they catch in flight at a rate of two insects per second.

Signals emitted by bats lie in the ultrasonic range (from 20 to 100 kHz), which accounts for a significant delay in the discovery of this phenomenon. Short wavelength of emitted sound makes it possible to receive echo signals from tiny objects. For example, a bat can distinguish a wire of 0.1 mm in diameter when it is pitch dark.

Using their ultrasonic sonar, some bats can hunt not only insects but fish that move close to the surface. For a long time it was unclear how they manage to do this since the acoustic impedance of the fish bodies is about the same as that of water. Therefore, fish should not be able to reflect acoustic signals. However, this is not absolutely true since the bodies of fish contain a swimming bladder filled with air. It is this organ that makes a fish visible to the ultrasonic sonar of a bat.

Bat echolocation calls consist of combinations of several audio components. The two most common are frequency modulated (FM) pulses and constant frequency (CF) pulses (Figure 6.9). FM sounds sweep through a range of frequencies in a very short time (usually in just one to five milliseconds). A typical FM echolocation call sweeps from 100 kHz down to 50 kHz in a span of only two milliseconds. The frequency of the sequence of such pulses can vary from 10 to 200 Hz, with the duration within the range from 0.5 to 5 ms.

CF-ultrasonic pulses last for more than 100 ms, and the frequency of vibrations remains constant during the entire pulse. The accuracy of maintaining the vibration frequency (about 0.05%) is truly phenomenal. Why do bats use such types of ultrasonic pulses for echolocation?

Judging by our experience, there is almost nothing that nature does for nothing. First, let us consider the principles of echolocation used by bats of the first type. Assume c is the speed of sound, and L_1 and L_2 are the distances from a bat to any two objects positioned on the way of the acoustic wave advance. Obviously, the pulse reflected by the first object returns to the bat in a period of time equal to $2L_1/c$, and the same pulse reflected from the second, farther, object returns in a time period equal to $2L_2/c$. If the frequency of sound in a pulse emitted by the bat were constant (CF), its duration higher than $2(L_2 - L_1)/c$, the pulses reflected from the objects would combine and give, as a result, an ultrasonic pulse of the same frequency but of somewhat longer duration (see Figure 6.10(b)).

Therefore, when using the CF-pulses, the reflected pulse would only contain information about the distance to the closest of the two objects when the distance between them were less than $c\tau/2$ with τ being the

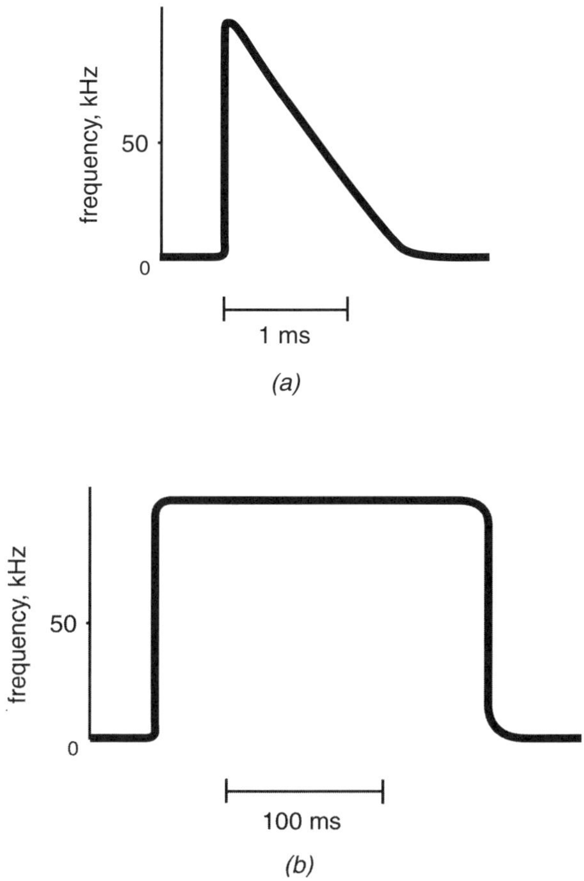

FIGURE 6.9. Schematic depiction of FM (a) and CF-ultrasound pulses (b) emitted by bats for echolocation.

duration of the pulse. Since vocal chords[1] of the bat don't allow for emission of acoustic impulses shorter than 0.5 ms, the bat will perceive the two objects located at a distance less than 10 cm as a single one. Can it really be that the image of the world in the bat's brain is so blurred? No, in fact, it does not happen.

The sharpness of the image significantly increases when the bat uses an

[1] Muscles of bat vocal chords are the quickest among all the muscles of all the known animal species.

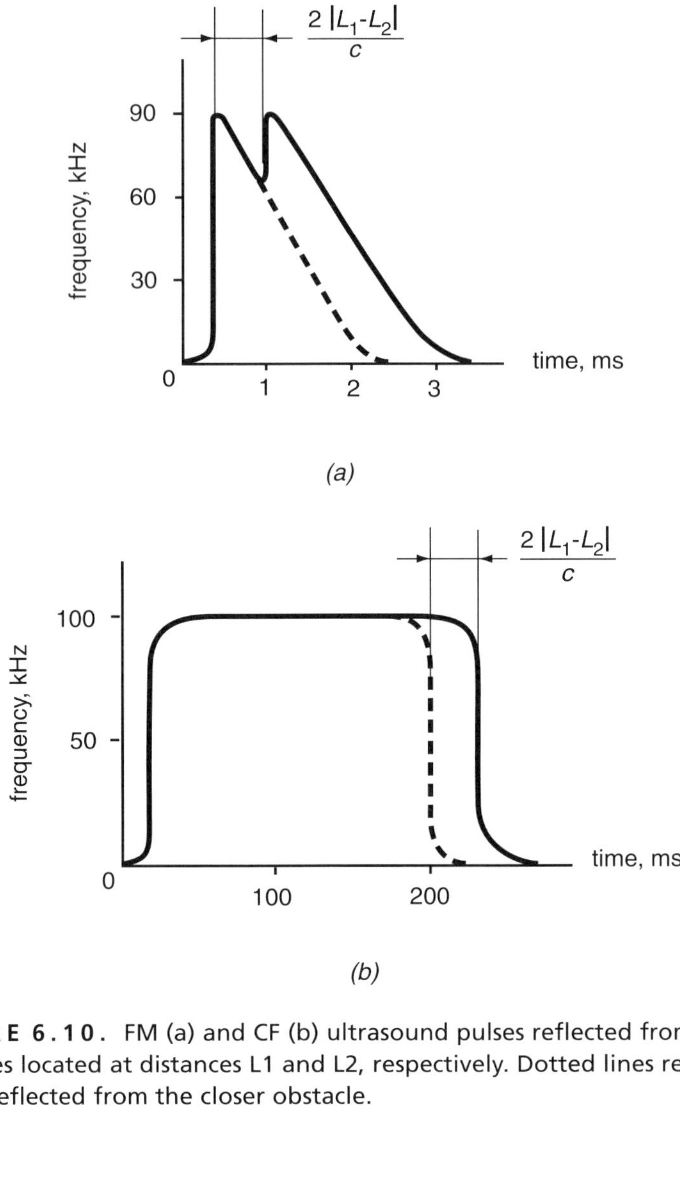

FIGURE 6.10. FM (a) and CF (b) ultrasound pulses reflected from two obstacles located at distances L1 and L2, respectively. Dotted lines represent pulses reflected from the closer obstacle.

FM-pulse. The FM-pulse reflects from two objects at a different distance from the bat and returns as an impulse the inner structure of which is shown in Figure 6.10(a). A sharp increase of the frequency of ultrasonic vibrations in a time interval $2(L_2 - L_1)/c$ after the arrival of the FM-pulse enables the bat to make a conclusion that two objects exist on the way of the acoustic wave. Obviously, unlike the case of the CF-pulses, in this case the duration of the emitted FM-pulse won't determine the sharpness of the echo-image.

Having justified nature's use of FM-pulses, let us now analyze the possibility of using CF-pulses to reach the same goals. It does not seem right even to refer to the CF-pulses as impulses since their duration sometimes can be greater than the interval between them. Such long impulses evidently can't be used for echolocation since the reflected signal would return to the bat before the emitted one terminates, which would impede the analysis of the reflected signal. Last but not least, the constant frequency of ultrasonic vibrations in these impulses would prevent bats from determining echo-signals reflected by different objects. Have we really found a defect in nature?

No, we haven't. That is just another illusion. We have forgotten that the bat uses its sonar in flight while moving at a speed of several meters per second. It is the movement of the bat that renders the echolocation system based on CF-pulses possible.

Suppose a bat moves relative to the surrounding objects at a speed v and emits ultrasonic vibrations with the frequency f_0. Then, due to the Doppler effect, the frequency of ultrasonic vibrations that reach motionless objects and reflect from them will be equal to $f_0 + \Delta f$, where Δf is defined as

$$\Delta f = f_0 v \cos\theta / (c - v \cos\theta)$$

where c is the speed of sound and θ is the angle between the vector v and a vector starting from the source of sound and ending at the point of observation. Due to the same Doppler effect, the reflected ultrasound is detected by the bat as vibrations with the frequency $f_0 + 2\Delta f$.

If we consider the average speed of the bat's flight to be about 5 m/s and the frequency of the emitted ultrasound 80 kHz, then the difference $2\Delta f$ between the frequencies of the emitted and reflected signals shouldn't exceed 3%, carrying information about the position of surrounding objects relative to the vector of the flying bat's speed.

To use this method of echolocation, the bat not only has to emit the ultrasound of a constant frequency (accurate to centipercent), but also to be able to distinguish the frequencies differing by 0.1%. Indeed, experiments showed that the acoustic organs of the bat have the highest sensitivity in the range of its own frequencies, so that they are able to react to a frequency change not exceeding 0.1%. Therefore, use of the Doppler effect and of pulse echolocation enable the bats to locate themselves without vision.

Nature took care not only of bats, having equipped them with ultrasonic sonars, but also of their victims. For example, some moths, common trophies of the bats, have an acoustic organ especially sensitive to the frequencies used by bats for echolocation. Butterflies defend themselves by ultrasonic signals jamming the operation of the bats' sonar and repelling the bats.

Praying mantis possess a very peculiar ear with the capability to detect echolocation signals of the bats (Yager and Hoy, 1986). Until recently these insects have been considered deaf. Then it was found that they do have a single hearing organ tuned to the ultrasonic frequency range from 25 to 45 kHz. The sensitivity of these insects to low-frequency acoustic vibrations (lower than 5 kHz) is decreased several hundred times. Thus, the praying mantis can be treated as an "acoustic Cyclops."

The only ear of a praying mantis is located in the middle of its small body and consists of two membranes at a distance less than 150 μm. Obviously, such a small ear size excludes the possibility of receiving any information about the direction of advance of the acoustic wave since the maximal delay for 150 μm will be less than 0.5 μsec, while the intensity of both membranes will always be the same, the distance between them being less than 1/50 of the wavelength.

Yager and Hoy came to the conclusion that, during the flight, the praying mantis constantly listens to the biolocation signals of the bats and when the signals reach a dangerous value, the legs and body of the insects tremble passionately, which results in the alteration of the flight direction. In such manner the acoustic Cyclops can avoid the chase of an insectivorous predator.

Vampire bats, who attack people as well as horses and other farm animals, rarely taste the blood of dogs. This is explained by the capability of dogs to hear ultrasound (the performance of dogs in the circus is somewhat based on this feature) and thus protect themselves when detecting a coming vampire.

Echolocation of bats and some cetaceans can help blind people. Biophysicists are already testing tiny earphone-shaped devices that constantly emit ultrasonic impulses and receive echo-signals reflected from surrounding objects. A special element of this device transforms the received echo-signals into an audible frequency range and then sends them to the earphones. After some training, a blind person equipped with such an ultrasonic sonar becomes somewhat capable of "seeing with the eyes closed." Many of the details of echolocation are not completely understood. Research on echolocation continues. For reviews, see an excellent book, *Hearing by Bats*, edited by Popper and Fay (1995).

7

Bone

We are all products of evolution. Nature had experimented for millions of years before it made us the way we are today. Unfortunately, we are not in a position to judge the results of this experiment in the sphere of human intelligence since, to evaluate human intelligence, we need the opinion of other rational beings, no contacts with whom have been established so far. At the same time, we can be reasonably impartial when we discuss the elements of mechanical structure of our bodies, comparing their properties to the parameters of similar elements used in engineering and construction.

A skeleton is the frame of the body, consisting of about 200 bones. Most of them (except for the cranium and the pelvic bones) are joined together in such a way that, during movement, their relative positions can change. Bones are driven by skeletal muscles, each of which is attached to two different bones. When a muscle is excited, its length decreases and the angle between the respective bones changes.

Figure 7.1 displays the conditions for the simplest problem of biological mechanics. From the given values of a shoulder and a forearm length and the weight of the load, we have to find the force developed by the muscle. Leonardo da Vinci was the first to formulate this problem. The solution of da Vinci's problem can be obtained from the equality of the moments of the

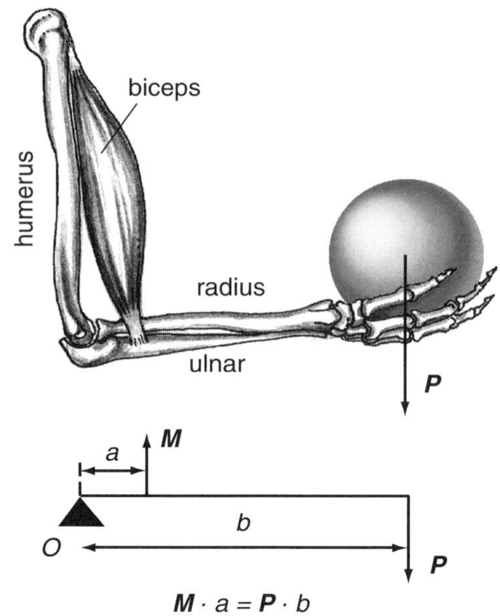

FIGURE 7.1. Holding a load *P* by a hand. (*M* = force of biceps.)

muscle force *M* and weight *P* relative to the pivot *O*. However, many other problems concerning the mechanics of the human body haven't been comprehensively resolved.

If mechanical engineers were faced with the task of designing a human skeleton, they would undoubtedly require that the purpose of each bone be explained to them. Indeed, the shape, size, and inner structure of a bone are defined by its function in a skeleton. Then, how do our bones operate? Like any construction elements, human skeleton bones are engaged in either compression and tension or bending. These modes of operation present nonidentical requirements to the skeleton bones. Everyone understands that it is difficult to break a match or a straw by tearing them along the axis, but they can be broken easily by bending.

In most cases, in both construction designs and animals skeletons, it is desirable for strength and lightness to be combined. How can we achieve the maximum strength in a construction given its mass and the strength of the material? The task is quite trivial when a construction element has to function either solely in longitudinal tensile stress or exclusively in compression. Thus, suppose that we have to hang up a load on a cable with a given length. The strength of the cable will be equal to the strength of its thinnest section; thus, its weight will be the lowest in the case of a uniform cross-section area along the whole length. The shape of the cross section itself plays no role as its strength is proportional to the cross-section area of the cable.

However, if the construction element is also expected to function in bending (for example, see the ulnar bone in Figure 7.1), then the problem of determining the maximum strength for a given mass becomes more complicated. Suppose a horizontal rod with a square cross section rests on two supports positioned at a distance L from each other, and bears a certain load, $2F$ (Figure 7.2). This loading configuration is known as a four-point bending. In this case the bending resistance of the rod to a great extent depends on the shape of its cross section.

As shown in Figure 7.2, the rod bends under the action of the force in such a way that its upper layers compress and the lower ones distend. Meanwhile, in the center of the rod there is a layer (more exactly, a surface) whose length isn't altered when the rod bends (shown as a thin line in Figure 7.2). According to any engineering textbook on elementary strength of materials (e.g., Hibbeler, 1994), the linear variation in stress across the cross section of the rod is given by

$$\text{stress} = \pm My/L \qquad (7.1)$$

where M is the bending moment at the cross section, and y is the distance from the neutral surface. The $\pm$ sign is used to indicate that one surface of the rod is subjected to tensile stress ($+$) and the other to compression (–).

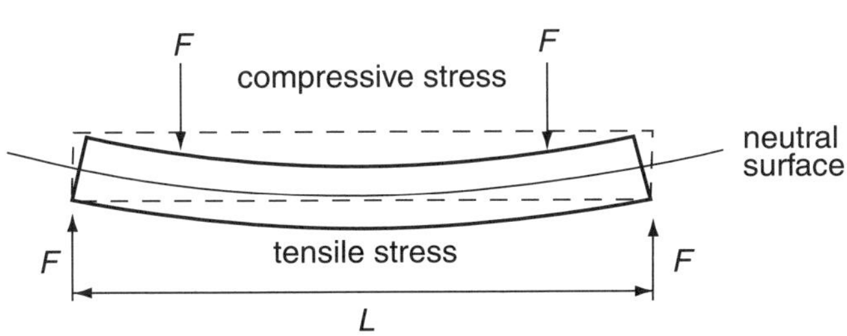

FIGURE 7.2. Four-point bending of a rod. See the position of the neutral surface in the rod.

Notice that the maximum stresses are experienced by the material on the rod surface (where y is the largest).

The material in the neutral layer doesn't perform any work (i.e., doesn't deform). It only makes the rod heavier. Therefore, the part of the material close to this neutral layer can be removed without decreasing the strength of the rod that operates under the specified conditions. However, this solution can be accepted solely for the rods with a square cross section. As for skeletal bones, as a rule, they have a circular (or near-circular) cross section. Proceeding from these considerations, an ideally constructed bone will have a partially absent core since the cylindrical layer around the axis of a bone doesn't undergo significant deformations under bending, but just increases the mass of the bone.

It is not surprising that in the course of evolution nature resorted to this way of decreasing the mass of human beings and animals, while preserving the strength of the skeleton. The effect of nature's design is especially pronounced in birds, whose interest in decreasing their mass is greater than that in other animals. This was first noticed by Giovanni Borelli, an Italian physicist, who wrote in his De motu animalium (1680) that "the body of a bird is disproportionately light compared to that of a human or any other quadruped, ... since the birds' bones are porous, vacuous, with the thinnest possible wall." Thus, a frigate bird, whose wingspan is about two meters wide, has a skeleton mass of only 110 g. Meanwhile, even in wingless animals, bones are empty inside. For example, the measurements show that for a femoral, the largest long bone of the skeleton, the ratio of the inner cross-section diameter to the outer one is equal to about 0.5–0.6 for a fox, lion, or giraffe. It enables all the animals (including us, of course) to decrease skeleton mass by about 25%, while possessing the same strength.

Composite Structure of Bone

The bone is a composite material consisting of two completely different components: an organic matrix and an inorganic mineral component. The organic matrix primarily consists of 90% collagen and 10% ground substance (mostly, glycosaminoglycans and glycoproteins). Calcium phosphate (called hydroxyapatite, or $Ca_{10}(PO_4)_6(OH)_2$) forms the inorganic mineral component. This mineral takes the form of small crystals, which are arranged in an orderly pattern within the collagen network. Collagen in a bone is one of the major components of the connective tissue; our tendons consist mostly of this substance.

The hydroxyapatite crystals are stiff and brittle with a poor impact resistance, typical of ceramics. The collagen fibers are much more extensible, so that the combination of collagen and apatite endow a composite with properties different from and better than those of the individual constituents.

Calcium salts are the major mineral component of the bone. Calcium atoms take up 22% of the overall amount of atoms in the bone. Note that other tissues of the body (muscles, brain, blood, etc.) contain about 2–3% of calcium atoms. Calcium is the heaviest element among the elements present in our organism in considerable amounts. Therefore, calcium's predominant localization in bones makes them distinctly seen in x-ray check-ups.

Each component of the bone can be removed easily without changing the bone's shape. For instance, if a bone is kept in the 5% solution of acetic acid for a sufficiently long time, then all mineral components (including calcium salts) will be dissolved in the acid. The remaining bone will consist mostly of collagen. It will become elastic like a rubber binder, so that it can be bent into a circle. By contrast, if the bone is burned, all the collagen burns out while the inorganic component remains.

Compact versus Spongy Bone

There are two major forms of bone tissue: compact (or cortical) bone and spongy (or trabecular) bone. The location of these bone types in a typical long human bone is shown in Figure 7.3. Compact bone comprises the walls of diaphyses (the shaft of a long bone) and external surfaces of a bone. Spongy bone exists as a three-dimensional interconnected network of trabeculae, which orient themselves in the direction of the forces applied to the bone. The classification of bone tissue as compact or spongy is based on their density. The density of compact bone may range from 1.8 to 2.2 g/cm^3, and that of spongy bone may range from 1.5 to 1.9 g/ cm^3 (Duck, 1990).

Bone resorption and formation occur throughout the entire human life and comprise the process of bone *remodeling*. Osteoclasts are the cells that resorb bone, creating longitudinally oriented tubular channels (the

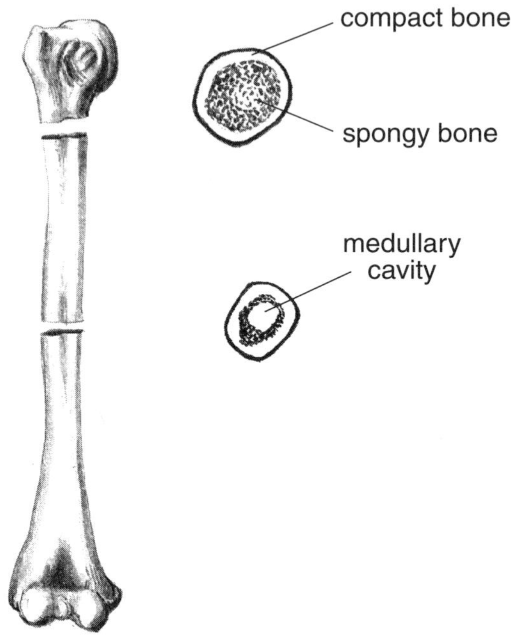

FIGURE 7.3. Structure of a typical long human bone showing the spongy bone, compact bone, and medullary channel.

Haversian channel). Osteoblasts, which deposit successive layers of lamellae on the surfaces of the Haversian channels, are the bone-forming cells. The Haversian channels communicate with each other and form the Haversian systems that help the osteocytes receive their nutrients as well as interconnect with each other and with the central hole (medullary cavity).

From a mechanical standpoint, a compact bone can be modeled as a cylinder structure consisting of numerous concentric layers with saturated bone liquid filling the space between them. In a complex bone structure, fluid flow causes considerable drag, imparting viscoelastic characteristics to the bone. It prevents the bone from instantaneously responding to an external load.

Spongy bone is a structural matrix of bony plates, beams, and struts with marrow in intervening spaces. Trabecular struts are typically 150–300 μm in diameter and form a three-dimensional lattice structure. Trabeculae align along the direction of a load applied to a bone, connecting to one another with struts. Therefore, the microstructure of a trabecular bone depends on the location and the direction of a load, which helps to optimize a strength-to-weight ratio in a healthy bone.

Trabecular bone has a branching pattern and exhibits self-similarity. That is, the trabeculae and the marrow spaces between them look very similar regardless of their size. Therefore, fractal analysis can be applied to the analysis of a trabecular bone, allowing for a noninvasive description of a trabecular bone microarchitecture (Lespessailles *et al.*, 1998; Fazzalari and Parkinson, 1998). Trabecular bone is generally defined as a cellular foam consisting of an interconnected network of rods and plates (Keaveny and Hayes, 1993). Mechanical behavior of foams can be described by analyzing the mechanisms by which the cells deform. The overall behavior depends on the type of cell structure, the volume fraction of solids (or relative density), and the properties of the cell wall material.

Bone Strength

Usually, elastic properties of bone tissue are measured with the help of mechanical testing, which involves the application of measurable force to bone specimen for which the changes in length are measured. Stress is equal to the applied force divided by the specimen's cross-sectional area, and strain is equal to the change in length divided by the original length. Stress-strain curve is plotted, on the basis of which the Young's modulus, or the slope of the curve in the linear region, is calculated. This is illustrated in Figure 7.4. The linear region is followed by a nonlinear region where "yielding" occurs. Yielding is followed by a nonelastic deformation, until, finally, a fracture takes place. The stress at which a bone material yields is called its *yield strength*, and the corresponding strain, its *yield strain*. The maximal stress and strain a material can sustain before a fracture are called its *ultimate strength* and *ultimate strain.*

The elastic properties of bone material vary for different individuals and for the anatomic position of a bone. As mechanical testing showed, compact bone is much stronger and stiffer than spongy bone. The corresponding Young's modulus for compact bone is between 4 and 27 GPa versus 1–11 GPa for spongy bone. Stylized stress-strain curves for compact and spongy bone in compression are shown in Figure 7.5. Notice a large difference between yield and ultimate strain for spongy bone. As a load is

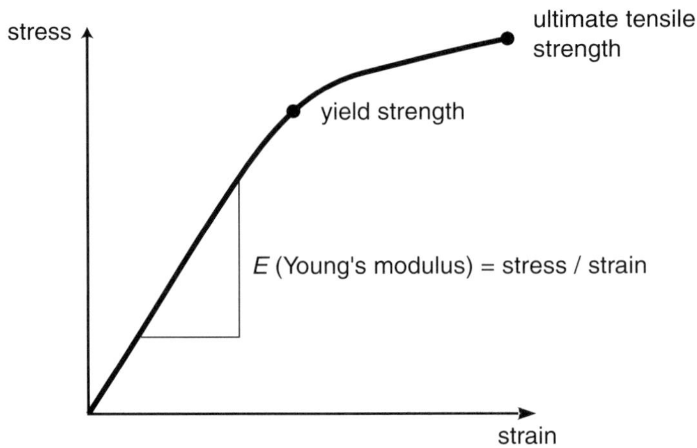

FIGURE 7.4. Stylized stress-strain curve for bone.

applied to a bone, the yielding of spongy bone is related to the gradual fracture of the trabeculae filling the marrow space. The long plateau region in stress-strain curve implies that spongy bone is able to absorb a large amount of energy upon impact.

Our bones have a large safety factor. The middle part of a human upper arm bone has the cross-section area of about $3.3\,\mathrm{cm}^2$. Using the data by

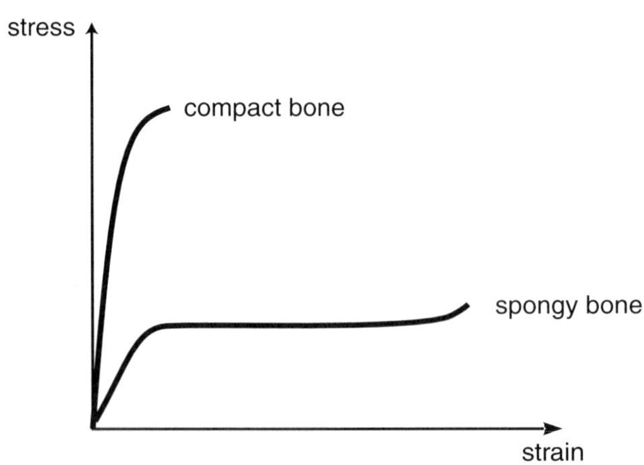

FIGURE 7.5. Stress-strain curves for compact and spongy bone in compression.

Reilly and Burstein (1975) for the ultimate compression strength of compact bone (193 MPa), we can easily show that the maximum weight of the load that can be held by this bone positioned vertically and operating under compression is close to 60 kN. At the same time, we can show that the maximum force that this bone can bear when the force is applied to its free edge perpendicular to the axis will be close to 5.5 kN (given the outer bone diameter of 28 mm, the inner bone diameter of 17 mm, and a length of 200 mm).

Why the bone is so strong is explained by its composite nature. Most common (noncomposite) materials that are very hard are also very brittle. Everyone knows how glass breaks. Cracks run from the point where the glass was struck, and they break the plate. When cracks have no time to form (for instance, when a bullet strikes glass), the glass plate remains intact except for the point where the bullet struck.

Thus, most materials could be much stronger if their structure prevented cracks from appearing and spreading. Highly elastic bone collagen presents such a barrier, preventing the cracks from spreading. On the other hand, the hardness of the bone is secured by mineral crystals deposited on the surface of collagen fibers.

Another material property of the bone is anisotropy. Therefore, its mechanical properties depend on the orientation of the bone structure with respect to the direction of force application. For example, the ultimate strength for a femoral compact bone tensed in longitudinal direction is 133 MPa versus 51 MPa when tensed in transverse direction (Reilly and Burstein, 1975).

Bone apparent density (i.e., the mass of a bone specimen divided by its volume) has a profound influence on the mechanical properties of bone. According to Gibson and Ashby (1988), Young's modulus (E) can be related to the apparent density (d) of the material by a power-law relationship, with n ranging from about 1 to 3, depending on the trabecular structure and the direction of testing:

$$E = a + b \cdot d^n \tag{7.2}$$

For spongy bone tissue, the ultimate compressive strength has been found to be linearly related to Young's modulus (Goldstein, 1987). Therefore, it should be noted that, as a result of the power law function, a small change in bone density could lead to a large change in stiffness and strength.

Of course, bone density is strongly related to the breaking strength. However, among patients with the same bone density, wide biological variation of ultimate strength can be observed. Therefore, researchers have sought to create more refined models of bone strength. The idea that the fractal index of trabecular bone might be related to bone strength is an appealing one, since the fractal index is easy to calculate from clinical images of a given bone, with the help of quantitative computed tomography. Measurements of a normal bone fall somewhere between 1.7 and 1.8. Fractal parameters derived from the trabecular bone texture analysis showed significant correlations with bone ultimate strength, so, in all hope, the fractal index may prove useful in estimating bone strength. Thus, fractal analysis of texture could provide a noninvasive technique of assessing ultimate strength and bone microarchitecture.

Osteoporosis

Osteoporosis is a disease that weakens our bones through the loss of bone density. Bones become weaker and weaker as bone density is lost, and osteoporosis is diagnosed when the bone density drops two or more standard deviation below normal, indicating loss of 25% or more of the total bone mass. Osteoporosis is manifested in fractures that result from any minor trauma. Every year in the United States, more than 1.4 million fractures that happen to people over 45 years old are attributed to osteoporosis.

In recent years, considerable effort has been taken in the research of osteoporosis. Investigators have been focusing on developing quantitative techniques to assess a human skeleton. The trabecular bone has a high surface-to-volume ratio, and a presumed turnover rate approximately eight times higher than that of the cortical bone. Therefore, the trabecular bone is highly sensitive to various stimuli. In keeping with these data, the clinical and epidemiological observations show that osteoporotic fractures initially occur in areas composed of predominantly trabecular bone, making it a prime site for detecting early bone loss and monitoring response to therapeutic interventions.

Osteoporotic fracture risk is currently predicted with the help of bone densitometry techniques. The most common technique for assaying bone density is based on the measurement of the x-ray attenuation coefficient with either a projection technique or its tomographic analogue, quantitative computed tomography. Whereas these techniques are capable of providing bone mineral densities, they do not give information on trabecular microstructure in terms of geometry, thickness, orientation, and density of the trabecular plates. All these data are usually obtained by optical stereology when biopsy specimens are analyzed microscopically. Magnetic resonance imaging has recently been applied to the study of osteoporosis as it has the potential to obtain information pertaining to both trabecular bone density and structure (Majumdar and Genant, 1995).

Wolff's Law and Bone Remodeling

For thousands of years bone has been used to make different musical instruments. No wonder: bone is highly stable to outer impacts. However, inside a living organism bone is surprisingly changeable since it is a living tissue of an organism. During our lifetimes, old bone cells die and new ones replace them. This is most noticeable during the first thirty years of life, when skeletal bones keep growing.

Every year, almost 500,000 patients worldwide get hip implants; about as many need bone reconstruction because of injuries or congenital defects. Long-lasting implants, which can be attached to a bone permanently, can save many patients from the pain, risk, and expense of undergoing multiple surgeries. Now, scientists are researching new technologies and materials that, when combined, promise to make bone implants almost as good as nature's own version.

Observing how bone is formed, scientists developed a method of creating pores and channels in metal implants and coating solid implants with these porous metals. As a result, new bone grows into these pores or voids to form a "lock and key" effect that attaches the implant firmly to the surrounding bone. Researchers from NASA (Innovation, 1993) are working with physicians to decide whether the space shuttle's ceramic surface insulation materials can be used as an implant for skeletal reconstruction. Theoretically, the idea of using space shuttle insulation is so attractive because such insulation is biocompatible and provides porous framework, which allows for infiltration by normal bone cells and for deposition of bone mineral.

In 1892, J. Wolff presented the first evidence that bone remodels in response to mechanical forces acting upon it. Since then it has been well known that bone grows where it is loaded and resolves where there is no loading. The objective of bone remodeling, along with the changes in bone mass, is to minimize flexural deformity of bone through a drift in the mass of bone towards the concavity created by the bending.

It has been demonstrated that the patients who lie motionless all the time lose about 0.5 g of calcium a day, which indicates the decrease of their bone mass. During the first space flights, due to weightless conditions, astronauts lost up to 3 g of calcium per day, which made most specialists doubt whether it was possible for people to participate in long-term space flights. However, special courses of physical training were developed, creating necessary loads for bone tissue, which significantly decreased calcium losses in weightless conditions.

Two general issues concerning bone remodeling have been debated for quite some time. What is the driving force behind remodeling? What is the mechanism that transfers mechanical signals to bone cells to provide remodeling? A variety of stimuli for bone remodeling have been suggested. For example, alteration of blood pressure has been shown to affect bone remodeling. Specifically, Kelly and Bronk (1990) demonstrated that increasing the venous pressure in bone, entailing an increase in the interstitial flow pressure gradient, causes bone deposition.

Bone is a piezoelectric material that generates electrical signals in response to dynamic mechanical loading (Gross and Williams, 1982). In principle, these electrical signals can trigger bone cells to control remodeling. This hypothesis was supported by experiments on animals, where externally applied electrical fields have been shown to affect bone remodeling. It has been suggested that experimentally observed strain-generated potentials in bone may stimulate bone remodeling. However, no definite experiment has been conducted so far to demonstrate that strain-related potentials actually affect or control bone remodeling (J. A. Spadaro, 1997).

Electrokinetic effects, which can really play a significant role in the process of bone remodeling, are due to the flow of fluid past an electrified solid-liquid interface. As fluid flows past the solid, it drags along a charge in the fluid at the interface, producing an electric current called a *streaming current*. The magnitude of the current depends on the magnitude of the charge imbalance and fluid velocity.

Figure 7.6 shows the direction of hypothetical electric field that emerges due to the streaming current in a bone under bending. The fluid would flow from the bottom surface (compressed) to the top surface (in tension). The fluid flow is responsible for the resulting electric field, whereby the distended surface of a bone always acquires a negative charge relative to the compressed one. If a bone is bent as shown in Figure 7.6 its concave surface charges positively and the convex one charges negatively. The corresponding tension of the electric field under standard loads usually doesn't exceed 0.5 V/cm.

It is well documented that the induction of electric current in a bone induces new bone formation (Yasuda, 1953). It turned out that if the electric current is passed through a bone for a long time, the mass of the bone substance increases close to the negative electrode. In this case, the required electric field tension is close to that arising under natural bend deformations. The mechanism by which electrical stimulation causes osteogenesis is still unknown, yet the related investigations go on (Yonemori *et al.*, 1996; Zhuang *et al.*, 1997).

Karate Mechanics in Short

What illustrates the strength of human bones best of all is the sport karate, so popular nowadays. Pictures of a karateist breaking tough wooden or

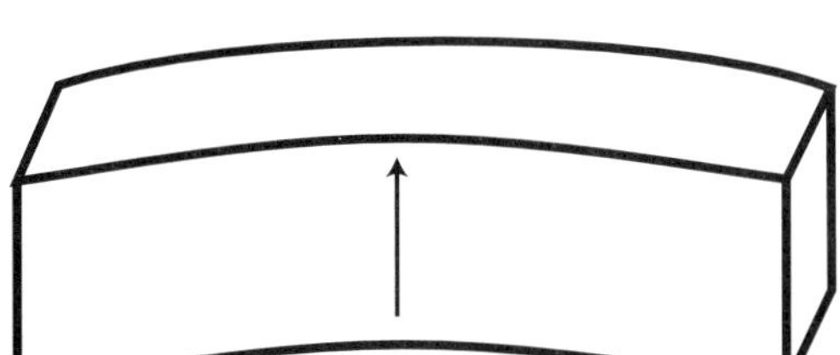

FIGURE 7.6. Direction of hypothesized electric field (arrow) that occurs due to the streaming current in a bone under four-point bending.

concrete blocks can be seen in lots of magazines and movies. Nevertheless, those who see it for the first time think it is mystification. But in karate even a newcomer, after a short training, can break a wooden block and then a pile of concrete blocks with a bare hand.

The techniques of Japanese karate style, which is widely practiced now, were developed on the Okinawa island. Having occupied the island in the seventeenth century, the Japanese disarmed the natives, forbidding the production and import of weapons. To protect themselves, the Okinawians worked out strategies of fighting with the help of a bare (kara) hand (te).

Karate methods differ from those of western weaponless self-defense. A western boxer transmits a large impulse to an entire body mass of the enemy, thus knocking him down, but a karateist concentrates his blow on a very small part of the enemy's body, trying to finish the blow at a depth of no more than 1 cm, his arms not swinging much. Therefore a blow of a karateist can easily destroy the enemy's tissues and bones at which it is directed. A well-trained karateist can transmit in a single blow the power of up to several kilowatts within few milliseconds.

The question is how a bare hand, without breaking itself, can break such tough objects as oak or concrete blocks? First, let us try to evaluate the energy W_b that is necessary to break a block. Applying Hooke's law to the block deformation and the formula for the potential energy stored in a compressed spring, we can obtain the expression for W_b:

$$W_b = (VT^2)/2E \qquad (7.3)$$

where V is the block volume; T, the ultimate strength in tension for the material and E, the Young's modulus.

Equation (7.3) confirms our intuition that the larger a block is, the more difficult it is to break. It follows from the same formula that the more elastic the material of the block is, the more difficult it is to break, since extending it requires more energy.

As a rule, during their shows, karateists use concrete blocks with the dimensions $0.4 \times 0.2 \times 0.05$ m. Keeping in mind that, for concrete, the ultimate strength in tension is 2 MPa, and $E = 16.5$ GPa, it follows from equation (7.3) that for such blocks, $W_b \approx 0.5$ J.

The speed of a karateist's moving hand is approximately 12 m/s, its mass being 0.7 kg. Hence the energy transferred by the hand at the moment of the blow is close to 50 J. Thus, the hand of a karateist has enough energy to break a concrete block.

The fact that the hand of a karateist doesn't break when it hits a concrete block can be explained party by bone being stronger than concrete. High-speed filming of a karateist's fist at the moment of the blow shows that its deceleration upon touching the block is about –4000 m/s^2. Therefore, the force acting from a block to the fist, the mass of which is 0.7 kg, must equal 2800 N.

If, at the moment of the blow, we replace the whole fist with a bone 6 cm long and 2 cm in diameter, fixed at two end points, and if we model the blow applied to a block with a force acting upon the middle of such bone, then in such conditions the bone can stand 25,000 N. This is approximately eight times greater than the force acting on the fist of a karateist when he breaks concrete blocks.

In fact, the ability of a karateist's hand to withstand such blows is even greater since, unlike a concrete block, it is not supported at the ends, and the blow isn't exactly directed towards the middle (see Walker, 1975). Besides, between a bone and a concrete block, there is always an elastic tissue that damps the blow. Therefore, we can't excuse our weakness saying our bones are brittle—they are not.

Leg Tendons—Living Springs

A lot of movements we make are periodic, such as walking, running, skiing, skating, squatting, etc. Doing these things, different parts of a body move in a nonuniform manner. For example, in the process of running or walking, each leg, in turn, gradually decreases its speed down to zero touching the ground and thus decelerating the movement of the body (Figure 7.7(a)). After that, the same leg pushes off the ground and accelerates this movement (Figure 7.7(c)).

To make a car move in the same manner, we would have to push, in turn, the accelerator and the brake with the frequency of about 1 Hz. Obviously, the fuel consumption in such an impulse mode of moving would increase greatly since a part of the kinetic energy of a car transforms into heat under braking. Is the running movement of a human or an animal as wasteful as the movement of this hypothetical car?

FIGURE 7.7. Direction of the force (shown by the arrow) acting from a runner to the ground at different stages of running: (a) braking phase, (b) uniform movement, (c) acceleration phase.

Of course, it isn't. The experiments were held with a testee who was made to run on a special tensimetric platform that was able to register all the components of the forces acting on him. From these data, supported by a simultaneous filming of running, it was possible to evaluate energy losses of the runner, assuming that the kinetic energy losses that occur at the deceleration phase fully transform into heat.

Additionally, actual energy losses could be calculated by measuring the rate of oxygen consumption by the runner since the energy amount liberated in the organism per 1 g of oxygen consumed by the body is well known. After both the estimates were obtained, it turned out that the actual energy losses in the course of running are 23 times less than those calculated on the base of the tensimetric tests (Alexander and Bennet-Clark, 1977; Ker *et al.*, 1987).

Thus, the assumption that all the kinetic energy in the braking phase transforms into heat is wrong. A part of this energy during *a-b* time range (Figure 7.7) is stored in the elastic tissues of legs in the form of the potential energy of their deformation, and during phase *c* it again transforms into the kinetic energy, as it happens when a rubber ball bounces off the wall.

The tissues that play the role of "springs" and that are able to store mechanical energy are limb muscles together with the sinews connecting them to bones. Compared to muscles, tendons are more suitable for storing the potential energy since the viscous forces inside them are very small, so that about 90% of this energy can be retransformed into kinetic energy.

Besides, tendons exhibit larger rigidity than muscles, and can be distended up to 6% of their initial length without being noticeably hurt, but muscles can be distended only up to 3%. All these characteristics of the sinews make them the main storage of the mechanical energy in the course of running and other cyclic movements.

The characteristics of the sinews are more or less similar for all the animals. However, the limbs of the ungulates, such as sheep or horses, are most adjusted to store the mechanical energy. Some muscles in the lower parts of these animals' legs consist of the tendons alone. The most impressive example of such use of tendons is the lower parts of a camel's legs, which are almost devoid of muscular fibers. The most powerful muscle in a human leg is the tendon of Achilles, which, in running, can experience a stretching force of up to 4,000 N.

We can verify for ourselves that the mechanical energy is really stored in our legs like it is in springs. To do this, just try to squat, energetically bending the knees. You will notice that it is much easier to stand up if you straighten your legs immediately, than after you keep your knees bent for a second or more.

The same tests were performed with the testees whose oxygen consumption was measured. In the first case, they were asked to sit down and, after fully bending their knees, stand up immediately, and in the second case, to stand up after lingering for a second and a half. The results confirmed the subjective observations: in the first case a testee consumed 22% less oxygen than in the second case. This can be accounted for by the fact that, in bending the knees, some of the muscles are stressed as they control the movement downwards, their sinews distended. If, before a person stands up, the sinews are prevented from shortening, the potential energy stored in them transforms into the kinetic one. However, if they are allowed to shorten before straightening up, then this energy will transform into heat.

8

Optics of the Eye

The human eye is the organ that gives us sight, which enables us to receive more information about the surrounding world than any of the other four senses. The human eye (Figure 8.1) is a wonderful instrument relying on refraction and lenses to form images. A human eye and a camera have many things in common.The retina, a membrane that lines the back of the eye, plays the role of film in a camera. It contains an array of photoreceptor cells called rods and cones that convert the light energy into electrical signals. These photoreceptor nerve cells react to the presence and intensity of light by sending nerve impulses to the brain via the optic nerve.

A diaphragm, called an iris (the colored part of the eye), automatically adjusts to control the amount of light entering the eye. The hole in the iris (the pupil) is black because no light is reflected through it, and little light is reflected back from the interior of the eye.

The lens focuses the light and creates a real, but inverted, image. The lens makes corresponding corrections for focusing at different distances. It is flexible and can change its shape and its focal length. This control is exercised by the ciliary muscles. The way the eye focuses light is interesting, because most of refraction that takes place is not done by the lens itself, but by the aqueous humor, a liquid on top of the lens. When light comes into the

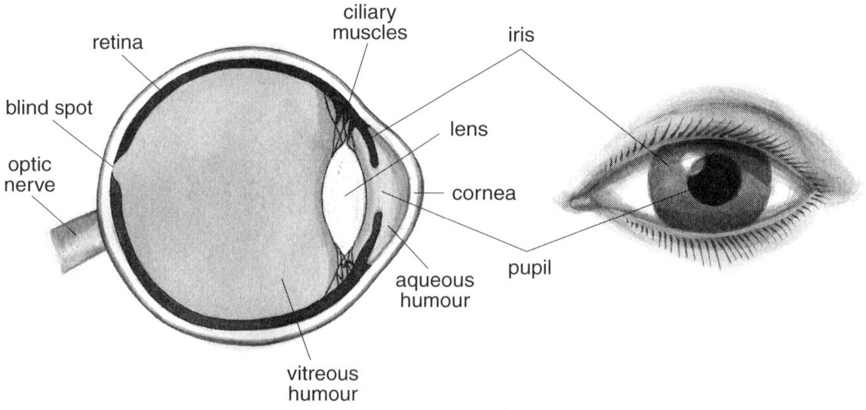

FIGURE 8.1. Structure of the eye.

eye, it first is refracted by this liquid, then refracted a little more by the lens, and then a bit more by the vitreous humor, the jelly-like substance that fills the space between the lens and the retina.

Johannes Kepler, a German astronomer, was the first one to guess that the image of the outer world forms on the retina. He came to this conclusion in 1604, before he discovered the main laws of the movement of heavenly bodies. His predecessors thought the eye lens was the organ sensitive to light. Hence, they all faced an insoluble problem: how can images of large objects get diminished in a tiny eyeball? To solve the problem, they had to assume that the lens was stimulated by only those light beams that fall perpendicular to its surface. Thus, they thought they simultaneously solved another problem: how the eye distinguishes between two beams falling to the same point of the lens from different points of the object.

Another contradiction that Kepler's predecessors faced and that seemed to be insoluble was the fact that the object image on the retina has to be inverted. To avoid this contradiction, they thought that there were regions in the eye that either prevented the beams from intersecting, or that birefracted them.

Kepler was the first to suggest that vision meant feeling the excitation of the retina by the inverted and diminished image of the object. He thought that each point of the object radiated not a single beam, but a solid cone of light. These light cones, coming from all the points of the object, entered the eye and, after being refracted by the eye lens, turned into the gathering light cones whose tops lie on the retina, creating images of the respective points.

Kepler also suggested that there was a very fine substance in the retina. In the points where the light beam fell on this substance, it was decomposed in a way similar to the decomposition of inflammables by light transmitted through a collector lens. He called this substance "visual spirit." Kepler's idea that the retina contains a certain substance that optically decomposes was confirmed only at the end of the nineteenth century.

Another prominent physicist contributed to the understanding of the operation of the eye as an optical instrument, E. Mariotte (1620–1684), one of the founders and first members of the Paris Academy of Sciences, who is also well known for his works on physics of gases and liquids. In 1666 he made a presentation at a session of the Academy devoted to the discovery of the "blind spot" or "Mariotte's spot" on the retina. The position in the back of the eye where the nerve (along with an artery and vein) enters the eye corresponds to the blind spot since there are no photoreceptors in this location. Thus, when the image of an object falls onto Mariotte's spot, the object becomes invisible. Mariotte repeated this experiment in the presence of the king and his courtiers and showed to them how to see one another headless. Normally, a person does not notice this blind spot since rapid movements of the eye and brain processing compensate for lack of information. This is the area that the ophthalmologist studies, examining a patient for glaucoma, a certain condition where the optic nerve becomes damaged mostly due to high pressure within the eye.

The quality of our vision is determined by the place where the light focuses in the eye. In a typical adult eye, total optical power (about 60 diopters) is well-matched to the eye length (16.5 mm), so distant objects are focused on the retina (*emmetropia*). There are a number of refracting surfaces in the eye, but the important ones are the anterior surface of the cornea, and the lens. Of these, the cornea, due to the large difference in refractive index between air (1.0) and corneal tissue (1.37), is more powerful, with a typical power of about 40 diopters. The lens in the relaxed (not accommodated) state has a power of about 17 diopters. Accommodation may increase it, by about 14 diopters with children, less with increasing age.

One common eyesight problem is near-sightedness, also known as *myopia*, in which light is focused in front of the retina. It blurs the vision of objects located at a long distance from the eye. A myopic eye is too long for its optics. In an *hyperopic* eye, far objects are focused behind the retina when accommodation is relaxed (i.e., the eye is too short for its optics). A hyperopic eye is farsighted because far objects are more in focus than near objects.

Visual acuity is the ability to see the details of an object separately and clear, and is the measure of the sensitivity of the visual system. It is expressed in Snellen notation as a fraction where the numerator indicates the test distance and the denominator denotes the distance at which a letter read by patient subtends 5 minutes of arc. Normal vision is expressed as 20/20 (or 6/6 in countries where metric measurements are used). Some of us may have a visual acuity of 20/10, which is much better than normal. It means that you can read letters at a distance of 20 feet, whereas most people can read only if they are within 10 feet from the chart. When a doctor says you have say, 20/50 vision in one eye, it means that this particular eye would need to be 20 feet away from an object to see as clearly as a normal eye at 50 feet away.

If normal human vision is 20/20, then a dog's vision is between 20/50 to 20/100, a horse's is 20/33, and a cat's is 20/100. However, it is difficult to measure acuity in animals, so studies have often shown wide variations in results.

Photoreceptors and Visual Pigments

The retina consists of about 130 million photoreceptor cells — the rods and cones of the eye (Figure 8.2). Rods are to intended to sense motion, and they work best at low light levels, allowing us to see in dim light at night. However, they do not give us any perception of color. Cone cells are adapted to see brighter light and can detect different colors. All mammals, including people, have more rods than cones. In humans, there are 20 times more rods than cones, found mostly at the edges of the retina. None are in the center, which is occupied by cones. This is why at night it is best to use peripheral or edge vision to try to see vague objects.

Rods and cones have stocks of membranous disks and an associated light-sensitive pigment. Rods and cones contain various visual pigments called chromoproteins (combinations of large proteins called opsins and a cartenoid pigment called retinene), which are activated by specific

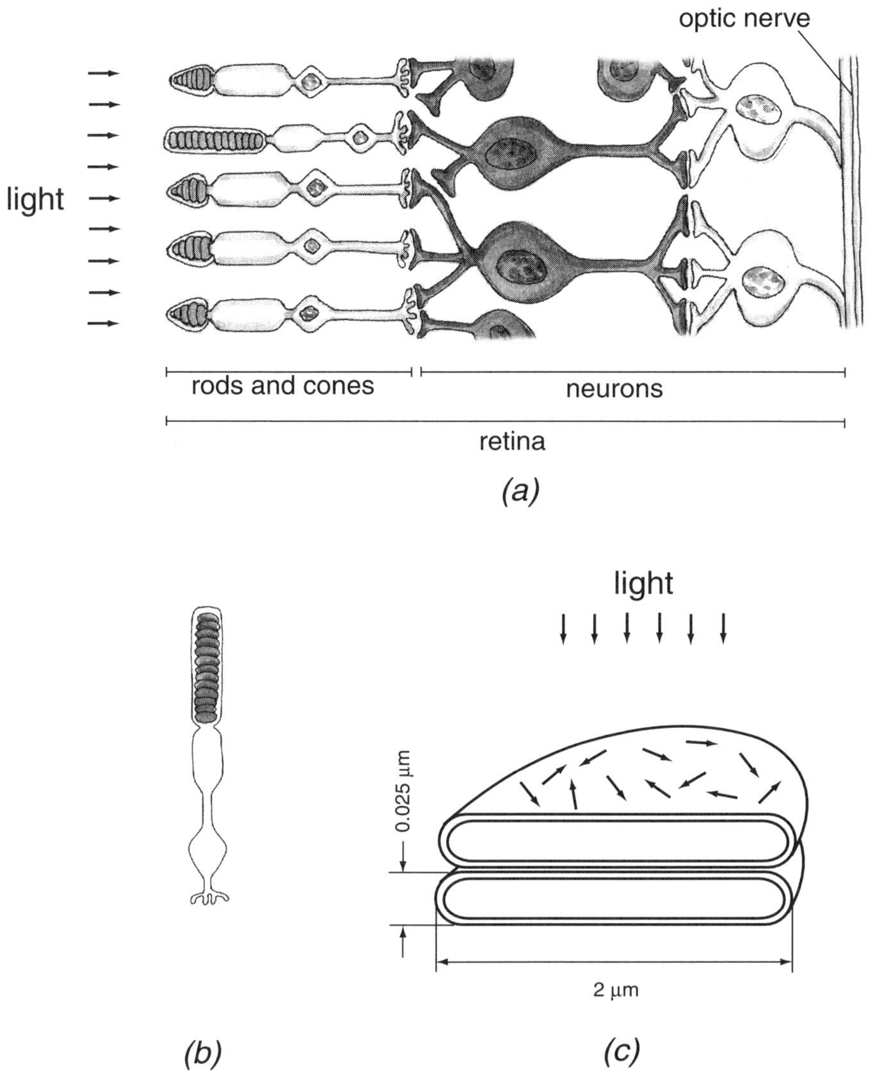

FIGURE 8.2. Structure of the retina (a), rod (b), and photoreceptor membranous disc (c) with arrows depicting the rhodopsin's dipoles.

wavelengths of light (Lythgoe, 1979). Rhodopsin is the light-sensitive pigment in vertebrate rod cells. Light-induced isomerization of the chromophore triggers a signal transduction cascade, which leads to closing Na^+ channels in the rod cell plasma membrane. Resulting hyperpolarization of the membrane is transmitted to connected nerve cells. The photoisomerization of rhodopsin in 200 femtoseconds is among the fastest and most efficient photochemical reactions known. The sensitivity of the visual pigment to light is so high that one photon is sufficient to induce the photoisomerization of rhodopsin (Baylor *et al.*, 1979). In the darkness, a reverse process occurs and rhodopsin again becomes sensitive to light.

Cones have visual pigments called photopsins that can detect bright colored light. There are three different types of cones with three different variants of photopsins, each particularly sensitive to either red, blue, or green light but all able to detect a wide range of colors. Dogs have two types of cones. Evidently a dog has a sort of vision similar to that of a person who is red-green color blind. Cats have three types of cones, like people, but do not have exactly the same color vision as we do. Colors that would appear very rich to us are more pastel-like to a cat.

The visual pigment is mainly intended to absorb the light. The denser the package of the rhodopsin molecules the more light a photoreceptor absorbs, and thus the better the eye's vision in the dark. The light falls on the outer segment of the rod, which is a cylinder surrounded by a membrane. The cylinder consists of several thousands of membrane discs densely packed in a pile (Figure 8.2(a), (b)). Each membrane disc is a closed flat sack formed by the membrane, which contains light-sensitive rhodopsin molecules. The packing density of these molecules is about $10^5/\mu m^2$ (Bownds, 1981). The optical density of this rhodopsin molecules package enables about 99% of the falling light to be absorbed by a layer of the substance only 40 μm thick. Simple estimations show that the concentration of rhodopsin molecules in the membrane of the photoreceptor discs is so high that it is practically equal to the physical limit for the particles with the molecular weight of about 50,000 D. It provides for a practically total absorption of the light by a layer of the visual pigment several microns thick.

The light is known to be an electromagnetic wave whose propagation speed vector, k, forms a triplet of mutually perpendicular vectors with the vectors of electric, E, and magnetic, B, field. The plane containing the E and k vectors used to be called the "plane of polarization" for the electromagnetic wave. Meanwhile, in most cases, the light that falls into our eye doesn't have a definite plane of polarization, because we often use light sources such as the sun or a simple filament lamp. Electromagnetic waves emitted by these sources have a constantly changing plane of polarization, so that the E vector can take all the possible positions inside the plane perpendicular to k.

A rhodopsin molecule absorbs photons as a result of the interaction of the electric field of the electromagnetic wave with the electric dipole of the molecule. A molecule's ability to absorb a photon depends on the value of the angle between vector E and the vector of the molecule's dipole moment. If the direction of vector E is parallel to the vector of the dipole moment of the rhodopsin molecule, the probability that a photon will be absorbed is the highest. On the contrary, if the E vector is perpendicular to the dipole axis, the molecule won't absorb the photon. For all other angles φ, between vector E and the dipole of the molecule, the probability that a photon will be absorbed apparently varies in direct proportion to cos φ.

Let's return to Figure 8.2(c). Giant rhodopsin molecules inside a visual rod form very thin layers ($\sim 0.005\,\mu$m) in photosensitive membranes, so that the vectors of electric dipoles of these molecules are usually parallel to the plane perpendicular to the direction of the light. This means that the vectors of the rhodopsin molecule dipoles lie parallel to the plane where vector E is. The position of the giant rhodopsin molecules in very thin flat membranes oriented perpendicular to the falling light strongly increases the probability for a photon to be absorbed, as compared to the case of a rhodopsin solution with the arbitrary orientation of the molecules. It can be shown that in the latter case about one third of the molecules wouldn't absorb light since their electric axes would be perpendicular to the direction of vector E.

Tapetum — Living Mirror

Many people must be scared when they suddenly see the phosphorescent eyes of a cat in the darkness. Anyone who has ever gone fishing, of course, remembers how he or she admired sparkling iridescent fish scales. Both the phosphorescent cat's eyes and the iridescent fish scales appear due to the fact that some biological tissues can reflect light.

Many animals have surfaces that reflect light perfectly. These surfaces, though not of metallic origin, reflect light like polished metal. In all these cases, reflection is the result of the interference of light waves on thin films. For the reflection from a thin film to be full, the two reflected waves must have a phase shift relative to each other, divisible by 2π radians. It should also be taken into account that when a light wave reflects from a medium with a higher optical density (i.e., with a larger value of refractive index, n), it changes its phase by π radians while the wave reflection by a medium with lesser optical density doesn't alter its phase.

Thus, when a light wave falls perpendicular to a film, to obtain maximum reflection the optical thickness of a film (the product of its geometrical thickness, d, and the refractive index, n) must be equal to

$$nd = k\lambda/4 \qquad (8.1)$$

where k is a natural number, and λ is the light wavelength. The minimal thickness of such film equals $\lambda/4$, and the ratio of the reflected light intensity to the falling light intensity is several percent (8% for a thin water film in air). When we deposit several consecutive thin films to any flat surface so that the values of optical density of adjacent films differ and their optical thickness is $\lambda/4$, we can make the value of the reflection index of the system close to unity if the number of such films is close to ten.

In engineering, multilayered mirrors for optical devices have been made since the end of the forties. Usually they are made by alternatively depositing thin films of magnesium fluoride ($n = 1.36$) and zinc sulfite ($n = 2.4$) to the surface of the future mirror, by way of condensing their vapors. Nature used combinations of the following pairs of materials to create mirror surfaces in the living organisms: water ($n = 1.34$) and guanine crystals ($n = 1.83$); air ($n = 1.0$) and chitin crystals ($n = 1.56$); and water and chitin crystals. What are these mirrors in the eyes of the animals for?

By the middle of the nineteenth century it had been well known that a human being can see because of the ability to focus with the help of a lens. It was at that time that people started to use artificial lenses for glasses. Mirrors also soon became an integral part of optical devices (telescopes, etc.). But before people started to use multilayered mirrors in engineering, it had generally been accepted that animals don't use the principle of a mirror for focusing since they would need polished metallic surfaces for that. Recently it has been proven that several organisms have eyes that operate according to the principles of mirror optics.

It is enough to look at the scheme of an eye of a scallop (Figure 8.3), a mussel living on the sea bottom, to understand that without a mirror this eye

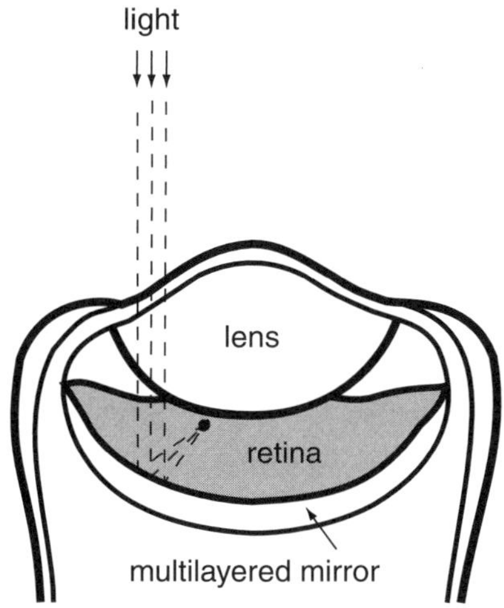

FIGURE 8.3. Schematic depiction of a scallop's eye.

would be absolutely blind. In fact, it seems surprising that the lens of this eye contacts the retina. It means that the lens creates a focused image behind the retina. There would not be a distinct image on the retina if not for a multilayered mirror, which returns the light beams to the retina and completes their focusing (Land, 1965).

Thus, the use of two focusing systems, a lens and a mirror, enables an eye to get a distinct image of an object with a significant decrease of the eye's volume. Besides, with such a focusing system the light passes through the retina twice, which immensely increases the eye's sensitivity. The latter feature of the eye seems to be most advantageous for deep-sea animals, which are forced to adapt to a very poor light.

Tapetum is an important structure of the eye that improves the night vision of numerous terrestrial animals. *Tapetum* is a peculiar flat mirror that makes a cat's eyes phosphoresce in the darkness. Like the spherical mirror in the eye of a scallop, the tapetum enables the part of light that passed through the retina without falling on the photoreceptors to return to the retina again. Partly due to that, the eye of a tiger can see surrounding objects under irradiance six times lower than is necessary for humans (Kitchener, 1991). Such a mirror has also been discovered in the eyes of some deep-sea raptors (Braekevelt, 1994 *a,b*).

Infrared Vision

Most animals see within the wavelength range of between 350 and 750 nm, a very narrow band in the spectrum of electromagnetic solar radiation. It may be so narrow because at shorter wavelengths the chromatic aberration, that is, the dependence of refractive index of the lens on frequency, becomes notable. Besides, high radiation frequencies are more intensely absorbed by the elements of the eye located in front of the retina. Longer wavelengths don't provide enough electromagnetic energy for photochemical reaction.

However, warm-blooded animals have one more obstacle to using long-wave radiation — heat radiation emitted by the body itself and penetrating an entire eye from inside. If the retina of warm-blooded animals were equally sensitive to the whole spectrum of the electromagnetic radiation, they would be absolutely blind since their heat (infrared) radiation would block light beams falling into eyes. The situation is completely different as far as cold-blooded animals are concerned. Here, the possibility for an infrared eye exists only if its sensitivity to the light is suppressed.

The feeling of temperature differs from all other feeling in that most of the animals, including humans, use it to provide for the sense of comfort rather than for orientation, finding food, or locating enemies. However, there are some exceptions: bugs, bees, mosquitoes, acarids, as well as asps, boas, rattlesnakes, and other reptiles. These animals use their thermoreceptors to get information about the objects very far from them. This is how some bloodsuckers find their future hosts. It has been found that bugs are able to find objects that have the temperature of warm-blooded animals at a distance of 15 cm.

Insects have sensitive thermoreceptors to register temperature signals coming from far away. These are located mostly in the antennas, special beards on the head, or on the legs, the latter serving to determine the temperature of the soil. Since insects have two antennas, they are able to determine exactly the direction towards the source of heat. Thus, having received the signal about the presence of a warm-blooded victim, a mosquito will keep changing the position of its body until both tiny three mm-long antennas register the same intensity of the infrared radiation.

Pit snakes have developed a remarkable type of vision that is extremely beneficial to nocturnal predation — the ability to see heat given off by warm-blooded animals. Infrared vision is accomplished by a combination of information from "pits" (small indentations around the mouth and eyes adapted to detect heat energy) and the normal eye. Impulses from both of those travel to the brain via the optic nerve. Here the images are combined to form a superaccurate image with which the snake can expertly determine the exact distance to a target prey even during the darkest of nights (Newman and Hartline, 1981).

Pits are usually positioned in front of and somewhat below the eyes of the snake, their number depending on the type of snake, and can be as much as 26 (python). Pits of a rattlesnake have been studied most thoroughly. The precision of the blindfolded rattlesnake's jump to a heat source (hot light bulb) is 5 degrees. The sensitivity of the pit is enough to detect a human hand or a living mouse at a distance of 0.5 m, which corresponds to the infrared intensity of about 0.1 mW/cm^2. Just for comparison, infrared intensity of solar radiation on a winter day at the latitude of New York City is close to 50 mW/cm^2. With the help of precise measurements it was shown that a snake jumps when the temperature of the sensitive membrane of the pit increases by only 0.003°C.

Why is a heat eye of a snake so sensitive, then? It is interesting that a snake has no special receptor cells sensitive to heat radiation. Humans also have nerve-endings of the same level of thermal sensitivity in the upper layer of the skin. But this is exactly where the solution lies: mammals (including humans) have these nerve-endings that are 0.3 mm deep inside the skin, whereas a snake has them inside the membrane that closes the lumen of a pit, the membrane being only 0.015 mm thick (Figure 8.4).

Although the upper layer of a mammal's naked skin is heated by the infrared radiation by the same value as the membrane of a snake, the sensitivity of the skin is 20 times lower. Besides, most of the heat energy dissipates in the mammal's tissues that don't contain thermoreceptors. By contrast, the pit of a rattlesnake contains some air on either side of the membrane, and air is known to be a good thermal insulator. Therefore the pit is so sensitive to infrared radiation not because of special thermoreceptors but due to the unusual structure of this organ.

The thermosensitive membrane of a snake appears to be an absolutely black body that absorbs all the falling light, while emitting only the part in which it is in thermal equilibrium. This obvious assumption makes it possible to evaluate the time of a snake's reaction to the appearance of the infrared radiation. The time constant, τ, is equal to

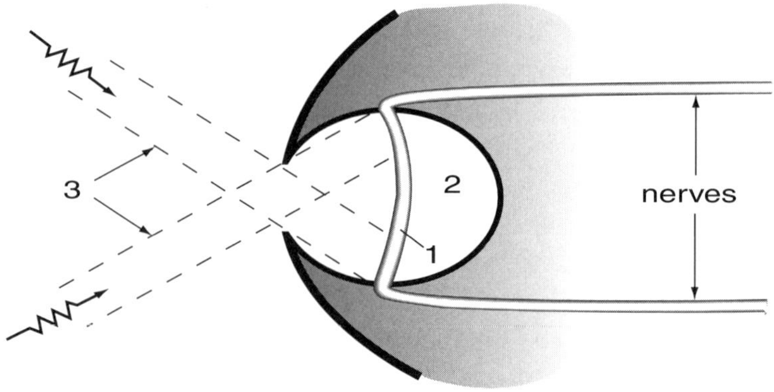

FIGURE 8.4. Section through the face pit of snake: 1) thermosensitive membrane; 2) air cavity; 3) heat radiation.

$$\tau = c/(8\sigma T^3) \tag{8.2}$$

with $\sigma = 5,7 \cdot 10^{-8}\mathrm{W/(m^2 \cdot K^4)}$; T is absolute temperature; c is the heat capacity of the unit area of the membrane. Taking the membrane's thickness equal to 0.000015 m and its heat capacity equal to the heat capacity of water ($4.8 \cdot 10^3\,\mathrm{J/(kg \cdot K)}$), we can calculate the heat capacity, c, of an area unit of the membrane that is equal to 72 $\mathrm{J/(m^2 \cdot K)}$. Substitution of this value into equation (8.2) gives the τ value close to 6 sec.

How does a snake determine the direction of the attack? The form of the pit is instrumental in this. As is obvious in Figure 8.4, the structure of the thermal eye resembles that of a pinhole-camera that was used to obtain images in the early years of photography. As a rule, the diameter of the thermosensitive membrane is twice as large as the diameter of the outer hole of the pit. However, the same figure shows that each pit possesses only primitive focusing ability. It enables the snake to distinguish between two separate infrared sources only if the angle between their directions is 30–60 degrees. At the same time, using several pits that have different overlapping views helps the snake to determine the direction to the object much more precisely after the brain has processed the data from all the thermoreceptors.

Compound Eyes

The arthropods (insects and crustaceans) consist of many segments (metameres) and are a huge phylum. Three times as many living arthropod species as all other animal groups have been described. Some kinds of arthropods have not changed in evolution. Dragonflies, for example, look the same as their early ancestors 300 million years ago, and could be called living fossils. There is an enormous number of insect species — some of them active only in the sunlight and others in both sunlight and shadow. There are insects who are active only in the dim light. Some of them (such as houseflies) fly at a great speed and make quick turns without slowing down, and others (such as dragonflies) fly solely along straight lines, and can soar in the air for a long time, turning on the spot.

Though they have different lifestyles, practically all insects have the same eye structure — the *compound eye* (Figure 8.5). The major difference between the compound eye and the human eye is their lens system. In the compound eye, the lens is divided into thousands of facets called *ommatidia* (over 10,000 in dragonflies). Each ommatidium is a separate eye that looks in its own direction. Therefore, each ommatidium transmits only a small portion of an overall image. Each ommatidium has its own lens that focuses the light onto several photoreceptor cells combined in a visual rod called a *rhabdome*. Acting upon a rhabdome, the light brings about a sequence of nervous impulses transmitted to the brain of an insect through the visual nerve.

Generally, there are two types of compound eyes — *apposition* and *superposition*. In some arthropods, *screening pigments* prevent the light from one ommatidium from falling onto the rhabdome of another, allowing a mosaic-like image to be formed (apposition eyes). In other species, screening pigments are absent. As a result, the compound eye of many nocturnal insects is characterized by ommatidia coordinated in a functional unity. In this case, light is utilized more effectively by a certain arrangement that allows incident rays from several lenses to reach the same ommatidium. This eye is called a superposition eye and the way it works is described in the next section. The so-called advanced orders of insect possess apposition compound eyes. In most cases, superposition eyes are found in insects that fly at night. When night insects with superposition eyes are forced to fly during the daytime, some of them can move screening pigments up and

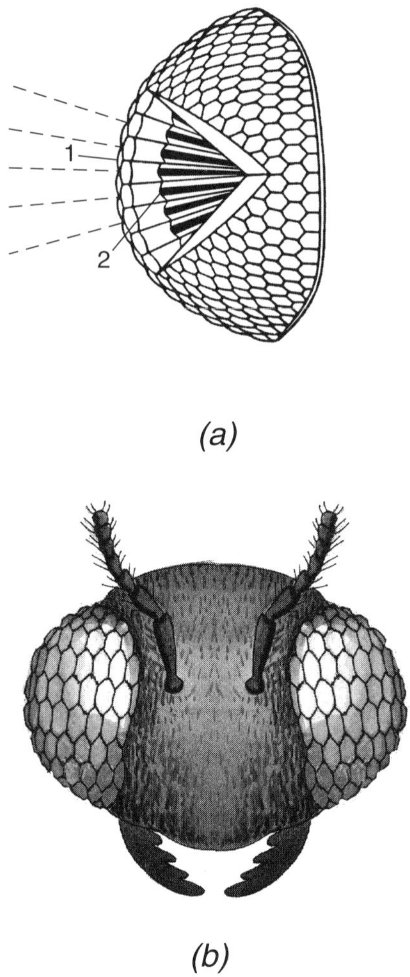

FIGURE 8.5. Housefly head (b) and structure of its compound eye: 1) lense; 2) rhabdome (a).

down the rhabdome, switching between apposition and superposition (Kunze, 1979).

Evidently, the main advantage of the compound eye is that this eye can see in all directions, whereas most mammals, including humans, have to turn the head to be able to look around. However, to get this advantage, an insect's eye had to sacrifice its resolving power. In fact, it is well known that the resolution of any optical system depends on the diameter of the hole through which the light enters the system. For the sake of simplicity we can assume the resolving power of the optical system (the minimal angular dimension of the object it is able to distinguish) is equal to $2\Delta\alpha = 2.4\lambda/D$, where λ is the light wavelength and D is the diameter of the optical system objective. As a rule, the diameter D of a single ommatidium is not more than 0.03 mm. It means that the resolving power of the ommatidium that operates in apposition mode is close to 1° of arc, given λ equal to 500 nm. For comparison, note that the diameter of a human pupil is approximately equal to 5 mm and provides for a greater resolving power of 200 times. That is the high price an insect has to pay for its "broad outlook on life."

Nevertheless, the resolving power of an insect's eye is quite enough for its owner. While flying above a book, an insect doesn't need to peer into the letters, as we do. The only goal of each ommatidium is to inform whether there is a bright object within its field of vision, and if, so, to determine its brightness. Ommatidium can't perform anything more than that anyway since it has only few photoreceptor cells with outgoing nerve fibers that send impulses to the brain with a frequency proportional to the brightness of the object.

What kind of structure must an insect's eye operating in an apposition mode have to locate a bright object most accurately? It seems reasonable that an insect's eye would function best if each light source corresponded to the stimulation of one or two adjacent ommatidia. The existence of several ommatidia that would look in the parallel direction and would thus be stimulated by the same point source of light, is evidently not an optimal eye structure, and would only increase its dimensions.

To narrow the field of vision of each ommatidium, it is enough to decrease the diameter, d, of its rhabdome and make it comparable to the dimension of its diffraction spot (Figure 8.6). Apparently, a rhabdome diameter smaller than that of the diffraction spot won't cause further narrowing of the field of vision of the ommatidium. Rather, it will just decrease its light sensitivity because the amount of photoreceptors in the rhabdome will decrease. Hence, the optimal diameter of the rhabdome of the ommatidium must be taken as $d \approx f\Delta\alpha$, where f is the focal length of the ommatidium lens. In this case, the minimum possible narrowness of the ommatidium field of vision is achieved, equal to $\Delta\alpha$ when its light sensitivity is maximum. In fact, here, as always, nature planned in advance,

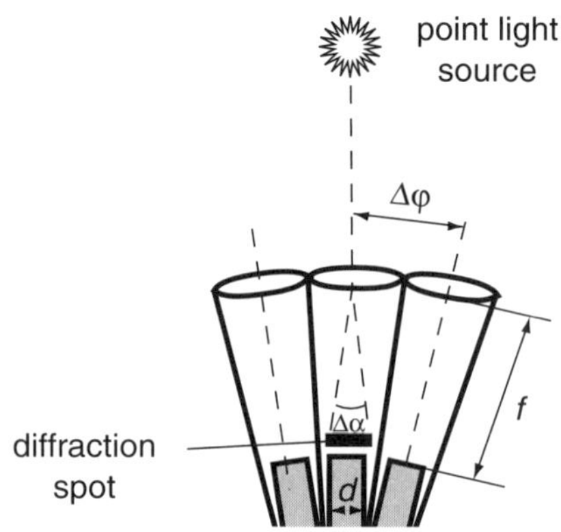

FIGURE 8.6. Analysis of optimal construction of a compound eye and the relationship between adjacent ommatidia (rhabdomes are dotted).

so that the actual ommatidium of an insect has the rhabdome diameter close in size to that of the diffraction spot.

However, to insure the sufficiently narrow field of vision of each ommatidium is only half of the solution. The eye structure must insure that, immediately upon leaving the visual field of the ommatidium, the light source can access the visual field of the adjacent ommatidium. To provide for that, it is sufficient to have the angle $\Delta\boldsymbol{\varphi}$ between the optical axes of the adjacent ommatidia close to the angular diameter, $\Delta\alpha$, of the field of vision of a single ommatidium; that is,

$$\Delta\varphi \approx \Delta\alpha = 1.2 \cdot \lambda / D \tag{8.3}$$

The analysis of geometric ratios in the eyes of different insects showed that relation (8.3) is valid for the species that are active in bright daylight. We can easily show that a hemisphere-shaped eye with the radius R (Figure 8.7) has a simple relationship between $\Delta\boldsymbol{\varphi}$, D, and R:

$$\Delta\varphi \approx D/R \tag{8.4}$$

Substituting (8.3) into (8.4) we obtain

$$D^2 \approx 1.2\lambda R \tag{8.5}$$

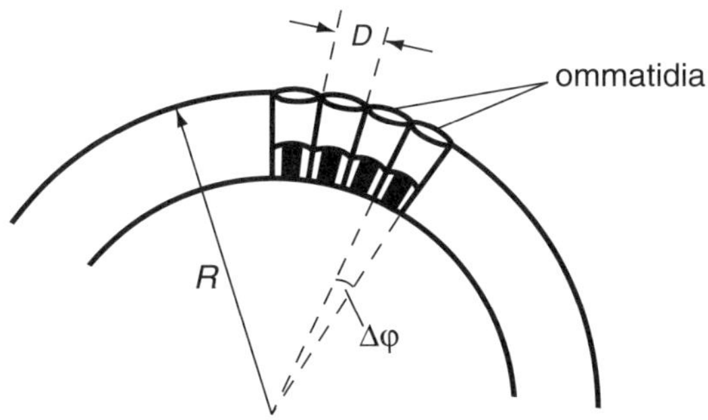

FIGURE 8.7. Helping to deduce formula (8.4).

Relation (8.5) shows that the size of the lens of each ommatidium, and thus the total amount of ommatidia, are unambiguously determined by the size of the entire eye. Using equation (8.5), we can now estimate D for a bee's eye that has the shape of a hemisphere with the radius equal to 1.2 mm. Taking $\lambda = 0.5\,\mu$m, we obtain $D = 27\,\mu$m, which coincides with the average data of morphometric measurements.

How Ommatidia Help One Another

One of the questions asked about the compound eye is whether this eye is always able to see. In fact, everything is determined by the light conditions in which a particular animal lives. In daylight insects (flies, bees, and dragonflies), a light beam that falls on each ommatidium is adequate to the sensitivity of the rhabdome. The compound eye of this particular structure performs its task beautifully, reproducing the space surrounding the insect. But if an insect is active only at night (like night-flies) or lives in dimly lighted places (for example, at the bottom of the sea), then a light beam passing through the ommatidia is not strong enough to stimulate the photoreceptors, and the eye becomes impotent. Therefore deep-sea shrimps, lobsters, and crabs, as well as night-flies, have faceted eyes operating in a different mode.

The operating modes of the eyes of different arthropoda species turned out to be different, depending on the extent of light in the place they inhabit. The mode we just described, the apposition mode, uses the ommatidia optically isolated from each other. In this case the lens of each ommatidium forms the inverted image of an object on the face of the corresponding rhabdom.

The compound eye can operate in the second mode — superposition. In this mode, the compound eye contains cooperating ommatidia. As a result, a direct (noninverted) image of an object appears on the layer of the rhabdome that is common for the whole eye. Unlike the apposition eye, the optical system of the ommatidium of the superposition eye is able to redirect the light to the face of the rhabdom of the adjacent ommatidium. Therefore, the optical and receptor systems of a superposition eye become common (for example, in a human eye).

Evidently, the apposition mode of the eye's operation is simpler, thus it can be found even in annulated worms. The superposition mode is much more sophisticated because it enables the eye to collect (focus) the light falling onto several ommatidia practically on a single rhabdome. It is generally accepted that the superposition eye appeared on the basis of the apposition eye, as a result of evolution as the distance between the layers of rhabdomes and ommatidium lenses gradually started increasing (Nilsson, 1983).

So how does the superposition facet eye operate? In 1891 S. Exner, a German biologist, wrote the book entitled *Physiology of Compound Eyes of Crabs and Insects*, which was considered revolutionary not only in biology, but in optics, as well. By that time it had become evident that the theory of light-beam refraction on spherical surfaces of common lenses can't explain the image focusing for many animals. The most complicated problem was faced by researchers studying the operation of the eye of crustaceans and water insects since the refraction index on the eye-water interface is relatively small, which makes the focal length much larger than the eye's size. Besides, it turned out that, for example, mantis have a practically flat outer surface of the eye.

Exner's approach was simple and brilliant. He asssumed that the optical system of each ommatidium is a cylindrical lens; that is, a cylinder made of some nonuniform material. The highest optical density is in the part of the cylinder along its axis, from where it falls towards the periphery according to the parabolic law (Figure 8.8).

We can easily show that optical structures similar to the one in Figure 8.8(a) focus light beams even though the surface they fall on is nearly flat. The difference between a cylindrical lens and a common one is that in the

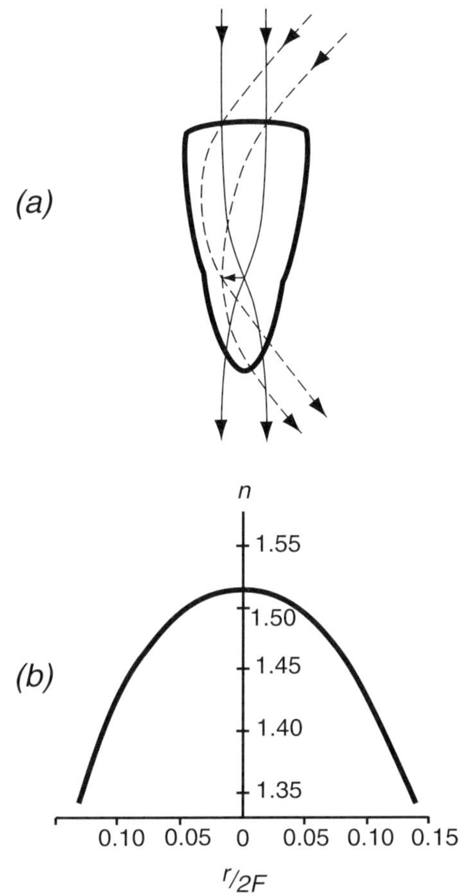

FIGURE 8.8. Cylindrical lens: the optical system of the night-fly's ommatidium: (a) route of beams through the ommatidium lens; (b) dependence of refraction index (*n*) on distance (*r*) to the axis of the lens expressed in relative units (*r*/2*F*), where *F* is a focal length of the lens. (Modified from Land, 1980. Reproduced by copyright permission of *Nature*.)

former an image focusing takes place in its bulk. Therefore, if the length of the cylinder is twice its focal length, then the parallel beams remain parallel after passing through the lens. Now, if they fall at an angle to the optical axis of the cylinder, then after passing through the cylinder, they will remain on the same side of the axis, retaining the same angle with it (see Figure 8.8(a)). It is this latter property of cylindrical lenses that turns them into focusing devices of the compound eye, devices that collect the parallel beam of light falling to different ommatidia on one rhabdome or several adjacent ones (Figure 8.9(a)).

Engineering had not exhibited any interest in nonhomogeneous cylindrical lenses until the end of the sixties. The interest that finally emerged, though, was not stimulated by the desire to make an analog of the superposition eye, but rather by an attempt to design optical fibers to transmit images.

However, not all insects or crustaceans have the optics of the compound eye based on cylindrical lenses. If we examine such an alternative compound eye under large magnification, it turns out that each ommatidium has a square crossection. Thorough measurements made it clear that the matter inside these ommatidia has the constant value of the refractive index, independent of the distance to the axis. How does this eye operate?

It was only in 1975 that K.Vogt, a German scientist, found the answer to this question. He assumed that the planes of the ommatidium crystal serve as mirrors for the falling beams. As a result, after reflection, each beam forms the same angle with the ommatidium axis as it did before the reflection, and remains on the same side of the axis. This is what provides for the necessary focusing of the beams (Figure 8.9(b)).

So a single reflection from the plane of the ommatidium crystal enables the facet eye to focus parallel beams. It is evident, though, that the focusing system shown in Figure 8.9(b) is able to focus only the beams lying in the plane of the drawing. To focus all other beams, a single reflection is not enough, whereas sequencing two reflections from the inner surfaces of the mirrors which form a direct angle solves our problem.

It is clear that, after the reflection from this pair of mirrors, the projection of the direction of the beam to the surface, perpendicular to their common edge, changes to the reverse one, which presents a necessary condition of focusing the image. Thus, the set of orthogonal mirrors in a superposition facet eye plays the same role as cylindrical lenses.

It is interesting that the principle of the facet eye operation we described was used immediately by astrophysicists to create x-ray telescopes. It is well known that x-rays are impossible to focus with the help of common lenses. That is why physicists highly appreciated the model proposed by biologists with regard to the operation of the eye of some cancroids. The first x-ray telescope based on the orthogonal system of mirrors was created in 1979.

To summarize the discussion concerning the superposition compound

(a)

(b)

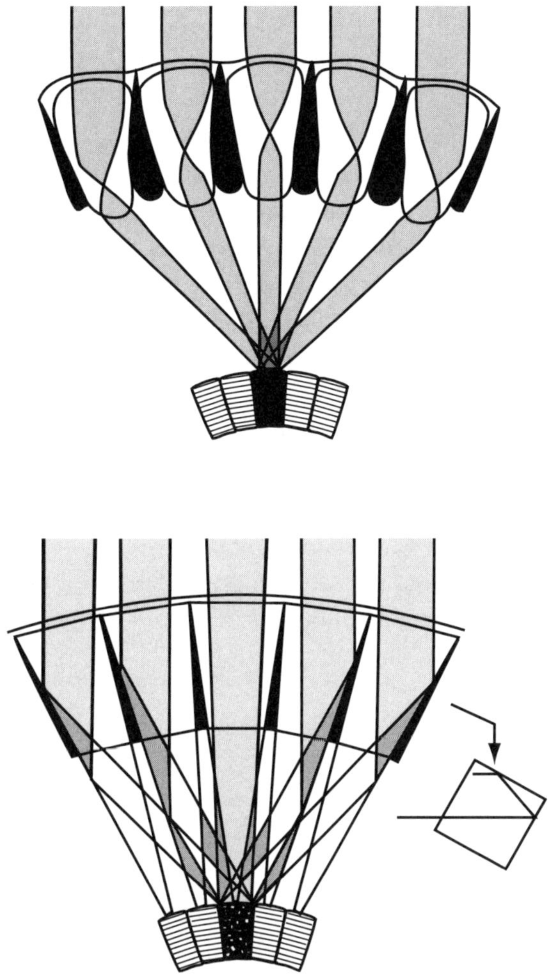

FIGURE 8.9. Schematic diagram of a compound eye forming a superposition focus of rays from a distant point-source: (a) the refracting superposition type employs a refractive index gradient in the crystalline cones, with highest optical density along the axis; (b) the reflecting superposition type relies on reflection and no lenses are necessary. In a top view (insert) the crystalline cones are square. Beams falling on an eye's surface are focused on a rhabdome (shaded) directed to the light source. (Modified from Nilsson, 1988. Reproduced by copyright permission of *Nature*.)

eye, I would like to draw your attention to the following. Such operation of each ommatidium is the necessary condition for focusing that makes it possible for the falling beams to be reflected from the plane passing through its optical axis. Three optical devices meet this requirement (Figure 8.10): (a) a system of two lenses or an equivalent cylindrical lens, (b) a flat mirror, and (c) a combination of a single lens and a parabolic mirror.

The first system is known to exist in compound eyes of night-flies and small shrimps, the second in the eyes of shrimps and lobsters. Finally, it was found that crabs and hermitcrabs have superposition facet eyes that operate according to the last of the mentioned mechanisms, characterized by the parabolic shape of the border between the crystal and the pigment that separates adjacent ommatidia. The surface of the parabolic mirror reflects a convergent beam that transforms into a parallel one, so that the angle between the falling and the reflected beam satisfies the necessary condition. The pigment lines the surface of the ommatidium crystal with layers the thickness of which is close to a quarter of a green light wavelength (about 90 nm), thus forming the multilayered mirror—a biological reflector with which we are already familiar (see *Tapetum—living mirrors*).

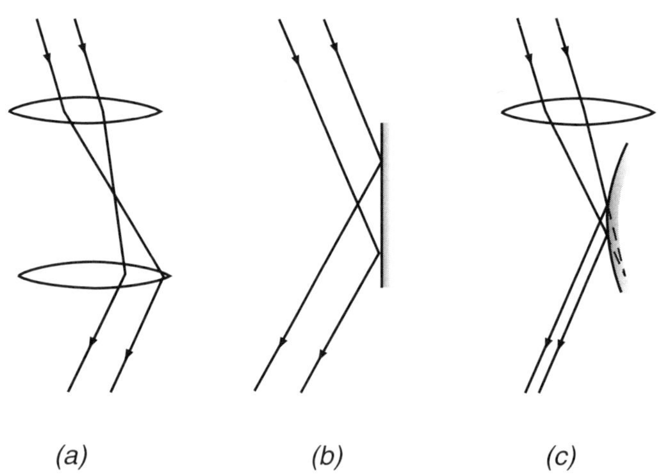

FIGURE 8.10. Three ways of inverting the direction of a light beam, corresponding to the three types of superposition eye: (a) refracting; (b) reflecting; (c) parabolic.

Microvilli See Polarized Light

At school we were all taught to find our bearings by the sun. But imagine that you are in the woods, and the sun can't be seen through the trees. It turns out that even in this case it is possible to determine the position of the sun in the sky. Unfortunately, a human doesn't possess this ability; therefore, we easily can get lost in the woods when we no longer see the sun. There is a large class of animals, though, who easily find the direction towards the sun even if they see but a small part of the sky. This class is insects.

Imagine you see an ant who is hurrying back home towards the anthill. You wait until its anthill is straight ahead, and then carry this ant about 50 meters to the right. The ant will keep moving in the same direction, finishing at 50 m to the right of the anthill. This is how it was concluded that an ant fixes in its memory not the spacial location of the anthill, but only its position relative to the sun. If, upon leaving the anthill, an ant saw the sun on the left at the direct angle to the direction of its movement, then on its way back home it must see the sun at the same angle on the right. The ant blindly obeys the orders of its compass.

Similar experiments can be performed on bees. It is easy to teach them to fly to a cup of sugar syrup. Moving the cup with the syrup from one place to another, it becomes evident that bees, like ants, find their way only relatively, according to the sun. In doing that, they do not actually have to see the sun itself: to find their way, it is enough for them to see only a small part of the blue sky.

Not only do the bees remember their way by the sun, they can also relate it to their neighbors in the hive. When a picker bee returns to the hive after a successful flight, it performs a very peculiar dance. The language of this dance was deciphered in 1945 by an Austrian zoologist Karl von Frisch (1886–1982). For his studies in the complex communication between insects, he was awarded with the 1973 Nobel Prize in physiology or medicine. An important implication of his work is that there is behavioral continuity between animal communication and human language (von Frisch, 1967).

By its dance, a bee informs its relations at what distance from the hive and in what direction the place rich in nectar is situated. Surrounded by its relatives, the bee moves along a "figure of eight" trajectory (Figure 8.11). The most informative element of the bee's dance is a straight line bisecting the eight (denoted by a curly line in Figure 8.11). Moving along this line, the bee shakes its body as if stressing the importance of that very element of the eight. The frequency of the bodily vibrations, as the bee is moving along the

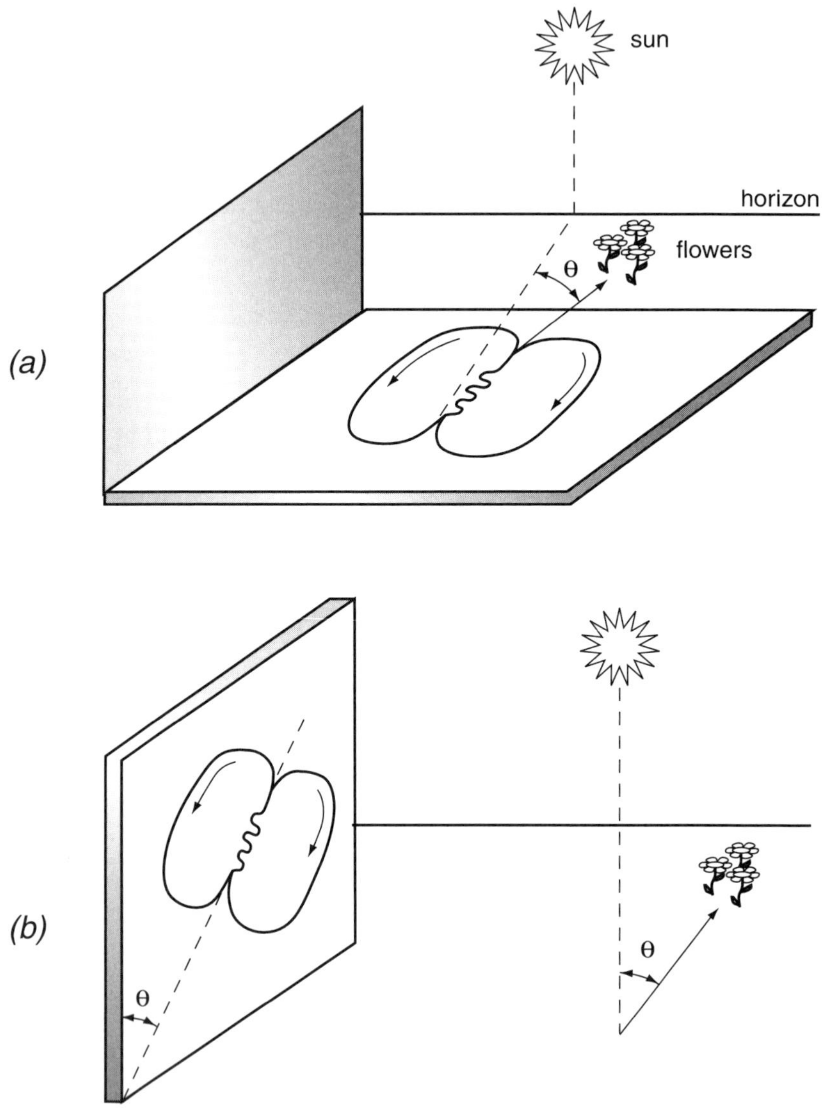

FIGURE 8.11. Dance of a bee on horizontal (a) and vertical (b) surfaces of the hive. It indicates the direction to the place rich in the nectar.

straight line, is about 15 Hz. At the same time, the bee also slightly buzzes with its wings.

If a bee performs its dance on a horizontal plane, the straight line of the bee's movement shows the direction to the place of the richest nectar collection. When the bee is dancing in the vertical plane inside the hive, it identifies the direction to the sun with a vertical, therefore the deviation of the straight line from the vertical will, in this case, be equivalent to the deviation of the way to the food from the direction to the sun. The time of the bee's movement along the straight line is in direct ratio to the distance between the hive and the place. For example, 1 s corresponds to 500 m, 2 s, to 2 km. In several minutes, the bees who have been watching the dance of their friend are already on their way to the specified place. The accuracy at which they reach the goal is about 20% of the actual distance (Wilson, 1972).

Experiments showed that when the insects can't see the sun, their unusual navigation abilities can be explained by the sensitivity of their eyes to the position of the polarization plane of the falling light (von Frisch, 1949). As it has been recently found, the similar sensitivity to the polarized light is present in some sea animals — some species of fish, shrimp, squid, and other cephalopods (Horvath, and Varju, 1995; Davies *et al.,* 1996; Flamarique and Hawryshyn, 1997). Remember that the eye of a mammal doesn't have such sensitivity because the electrical axes of the rhodopsin molecules are oriented randomly in the plane of the photoreceptor membrane (see Figure 8.2(c)).

The light emitted by the sun has no definite polarization plane. However, upon passing through the earth's atmosphere, the solar light is scattered on its molecules and particles, the dimensions of which are smaller than the wavelength of the light. As a result, each point in the sky above us turns into a secondary, partly polarized, light source, the polarization axis (the predominant E direction) always being perpendicular to the plane of the triangle with the tops formed by the observer, the sun, and the observed point in the sky.

It is also possible to solve the reverse problem: how to find the direction to the sun given the axes of light polarization from the two points observed in the sky. Obviously, the direction to the sun will be a straight line formed by the intersection of two planes, each containing the observer and being perpendicular to the corresponding axis of light polarization in a given point of the sky. This seems to be the way the insects find the direction to the sun, their eye being sensitive to the direction of light polarization.

What is it then, that makes the visual cell of an insect sensitive to the polarized light? Figure 8.12 shows a photoreceptor cell of an insect. Comparing Figures 8.2 and 8.12, the first thing that leaps to the eye is the different shape of the photoreceptor membranes. Photoreceptor membranes form flat discs in the visual rod of mammals, but in insects they are folded into long tubes (microvilli). Meanwhile, as tests showed, the electric axes of the rhodopsin molecules have the same direction over the whole photoreceptor cell of an insect. It is evident that, in this case, rhodopsin molecules can absorb only those photons in which the direction of vector *E* is parallel to the microvilli axis. Therefore, a simple transformation—

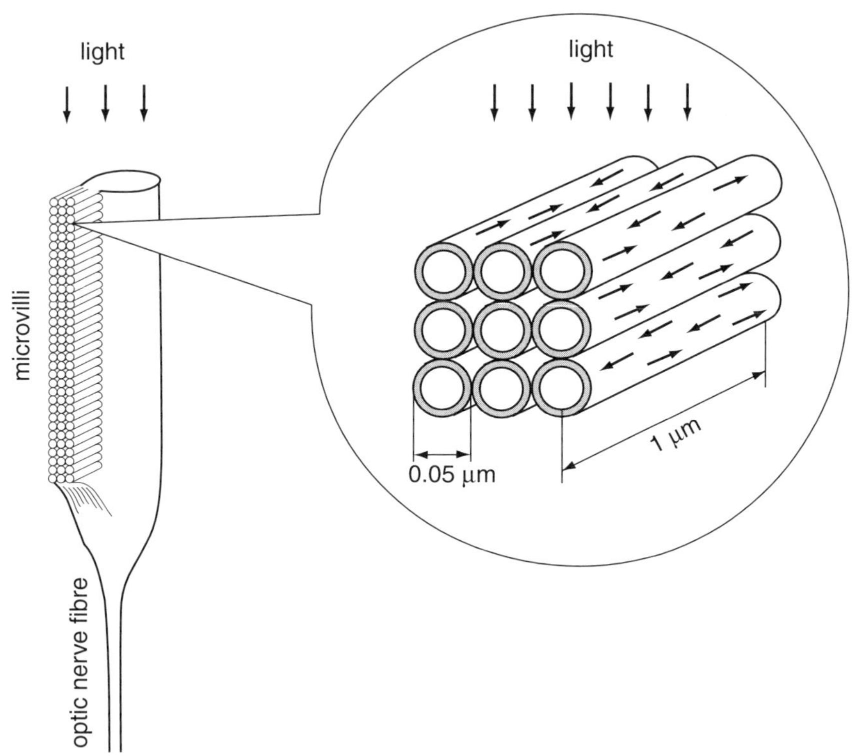

FIGURE 8.12. Schematic diagram showing the structure of the insect photoreceptor cell (left) and the arrangement of the photoreceptor membranes with the microvilli running in a single direction (right).

folding of a flat membrane into a tube—gives the photoreceptor a new quality: it becomes sensitive to the polarized light.

Animal Maps

It is not only research methods, but also the logic of research that biology and physics have in common. Experiments showing navigational capabilities of insects and birds are another example of how peculiar logic can be applied to the insects' behavior analysis.

J. M. Fabre was the first to try to solve the problem of how the insects find their way back home (to the hive, anthill, etc.), and the experiments he started 80 years ago are continued nowadays. Fabre was impressed by the capability of bees and wasps to return to their nests when they were taken to a distance of several kilometers away from home. Unable to provide any explanation, he introduced a concept of the instinct of direction. Later, analyzing the works of Fabre, scientists came to the conclusion that bees and wasps can remember the landscape that surrounds them. Moreover, flying insects, who can rise higher than the objects surrounding them, obtain much more information about what is around them than the insects who rarely see the circle of the horizon. Therefore, the insects who fly high (bees, wasps) usually find their bearings and the way home more easily.

An indirect proof that, to find their location, bees and wasps are governed not by an alleged directional instinct, but rather by the visual memory due to the panoramic view of the landscape, was contained in the experiment performed by G. J. Romanes more than a century ago (cited after Collett, 1987). A hive with bees was carried into a house several hundred meters from the seashore. The house was surrounded by flower gardens. The researcher started by opening all the windows in the house to enable the bees to study the surrounding landscape. In the evening of the same day he gathered all the bees who returned to the hive into a special box. On the following day, keeping the hive closed, he moved the box with the bees to the seashore at a distance of 250 m from the hive, and released the bees. Not a single bee returned to the hive from this "flowerless," and thus absolutely unfamiliar, place. By contrast, when Romanes performed a similar experiment, releasing the bees close to the flower gardens, all the bees found the hive in a moment, no matter at what side of the house the flower garden was. It means that, having found the nectar immediately beneath the windows of the house, bees would fly nowhere else and, therefore, would have no idea about the space surrounding the flower gardens.

Bees can find their bearing according to the map that they keep in mind. It was previously thought that only the vertebrata have this ability. According to the previous hypothesis, insects could remember only their old routes. The shortcoming of this concept was demonstrated by Gould (1986), who trained bees to find the food in an exact place at a distance of 160 m from the hive (point *A* in Figure 8.13). After the goal was reached, he marked the bees, collected them when they were leaving the hive, and moved them to a different place inside a dark box (point *B* in Figure 8.13) so that the hive, point *A*, and point *B* formed an equilateral triangle. It should be noted that point *B* was in a clearing in the forest, and point *A* was by the woodside. Therefore, the bees trained to find the food in point *A* couldn't see point *B* when they were flying.

After that the marked bees were released one by one, marking the direction in which they disappeared. It turned out that the average direction (among the 25 bees) in which the bees disappeared was 324 degrees to the north direction which was very close to the direction from A to B (330°). All the bees reached *A*, none of them flying towards the hive. If the "mind" of

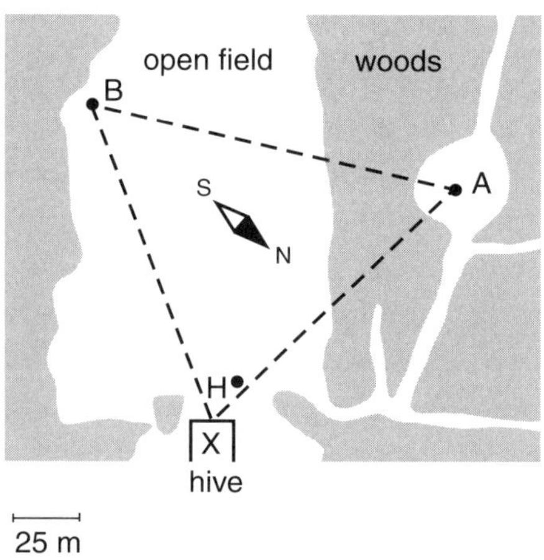

FIGURE 8.13. Plan view of test area: X, hive; A, foraging site; B and H, release sites. (Modified from Gould, 1986. Reproduced by copyright permission of *Science*.)

the bees had registered only the direction towards *A*, then they would have flown from *B* at an angle of 270 degrees. If they thought that they were already in *A* and flown towards the hive, then the direction of their flight would have been at an angle of 90 degrees. However, that is not what happened. The bees proved to have access to a sort of a landscape map, so that they could determine their position according to this map, and thus find the right direction toward point *A*.

9

Magnetic Sense

Though we don't feel any magnetic field around us, many animals react to its slightest changes. For example, wood mice can find their way in a forest following the direction of the magnetic field (August *et al.,* 1989). It was proven in the following way: A mouse caught in a park would immediately be placed into a special hermetic container equipped with two electromagnetic coils. The coils would be positioned so that it would be possible to reverse the direction of the magnetic field when the electric current was passed through them. Two minutes after the mouse was caught it would be taken in the container 40 meters to the north and then released; the direction of its movement for the next four minutes being watched. It turned out that, if the electromagnetic coils were de-energized while the mouse was being moved, the average direction of its movement after the release would coincide with the direction at the moment of the capture. But if the direction of the magnetic induction vector in the container was reversed, the released mouse would start moving in the opposite direction.

Not only does the magnetic field determine the navigational capabilities of mice, but it also influences their well-being. According to the data of Laforge *et al.* (1986), a group of mice placed inside a cylindrical container between the poles of an electric magnet were visibly alarmed after the magnetic field with the induction 0.9–275.5 mT was turned on (the magnetic field of the earth is 0.05 mT). Another sign of their reaction was that their weight gain changed. A group of testees used to gain 1g every day, whereas upon the magnetic field experiment the mice would first lose weight, and then they would "catch up," compensating for their weight losses.

On a Wing and a Vector

Intuitively, we can predict that magnetic sense must be of most benefit to the birds. In fact, during long migration flights, birds face numerous navigational problems since, due to the danger of raptor's attacks, the birds usually fly at night. Nasty weather doesn't prevent migrants from flying, either. However, it is not very easy to study magnetic sense of migrant birds as they use it only twice a year. A homing pigeon is a more convenient object to study magnetic sense. Taken tens or even hundreds kilometers away from its pigeon-house, it invariably finds the way home.

Installing a radio transmitter on the body of a bird, we can watch the entire route of a pigeon from the place it was released to the pigeon-house. It turned out that most of the time homing pigeons fly along a straight line as if they can see their pigeon-house hundreds of kilometers away. It is only during the first three minutes of the flight that pigeons decide which way to choose and so often change the direction of the flight. The same happens when they approach a point 2–5 km away from the pigeon-house.

The fact that choosing the right direction of the flight is not connected with homing pigeons' vision can be illustrated by tests in which birds' eyes are closed with special matte glass. After that, pigeons still managed to choose the right direction, though the only thing they could do was distinguish between day and night. Of course, such "blind" pigeons could not fly into the pigeon-house. They approached it from a distance of several kilometers and either stopped or started to fly randomly, changing places. Like migrants, homing pigeons can faultlessly find their way home at night or in nasty weather.

What kind of experiments finally convinced scientists that birds use their magnetic sense to find their bearings in long flights? In 1971 W. T. Keeton fixed a small magnet on the head of a homing pigeon. After that the pigeon was unable to find the way home, whereas another object of the same mass and size didn't prevent the bird from finding its pigeon-house. In the subsequent experiments, the head of the pigeon was equipped with a coil through which electric current could pass and thus alter the vertical component of the magnetic field of the earth. The latter is known to be directed downwards in the northern hemisphere and upwards in the southern hemisphere. These tests showed that such changes of the magnetic field change the flight direction of a pigeon by 180 degrees.

It is well known, however, that the magnetic field of the earth is subjected to significant fluctuations since it has two sources. The first one is the earth itself, itself being a huge magnet. The second one, which is relatively weak, presents the currents of charged particles (ions) in the upper layers of the atmosphere. Daily fluctuations of the atmospheric temperature lead to corresponding changes of ionic currents in the atmosphere. As a result, this component of the magnetic field of the earth experiences oscillations, from 3 to $6 \cdot 10^{-8}$ T in amplitude, which is about 0.1% of the average value of the magnetic field induction ($5 \cdot 10^{-5}$T).

Sometimes we face more significant changes of the magnetic field due to the increase of solar activity. A burst of solar activity leads to the increase of the ionic concentration in the atmosphere, which results in the increase of the corresponding component of the magnetic field. The amplitude of such sharp changes of the magnetic field of the earth, called magnetic storms, can be as much as 10^{-6} T. Experiments showed that during magnetic storms navigational capabilities of homing pigeons decrease dramatically. The birds face the same difficulties when, on their way back home, they encounter magnetic anomalies, that is, places where the induction of the magnetic field of the earth sharply changes for several kilometers, sometimes by more than 10^{-6} T. If the amplitude of such anomaly on the way of a pigeon exceeds $5 \cdot 10^{-6}$ T, the bird completely loses its navigational capability.

Bees are even more sensitive to magnetic field. Beekeepers know that wild bees build their combs in the same direction (with respect to the north-south line) as the one by which they were oriented in their native hive. If the new hive is put within a strong magnetic field, the orientation of the combs will be disturbed. It is thus accepted that the capability of the bees to sense the magnetic field of the earth enables them to coordinate construction in a new hive (Walker, 1997).

Another consequence of a bee's ability to sense changes of the magnetic induction is a well-defined schedule of the work in the hive. Indeed, how can these insects keep a 24-hour period of flying out, accurate to 15 minutes? Even when the weather is bad so that bees have to stay inside all day long, this rhythm does not change. However, once the magnetic field in the hive is changed, the "flying out" routine changes. It is generally accepted that the factor that coordinates the work of all bees in a hive is the earth's magnetic field, which changes every 24 hours. The highest rate of this change is at midday, and is greater than $3 \cdot 10^{-10}$ T/min. Such sharp changes of the magnetic field seem to signal all the working bees to fly out (Kirschvink *et al.*, 1992).

How can pigeons and bees measure magnetic induction? In engineering, the induction of magnetic field is measured mostly by two methods. In the first one, the electromotive force is measured between the edges of a conductor that moves in the field. In the second one, the measured quantity is the mechanical moment with which the field rotates a permanent magnet parallel to its lines of force. The first method is used by sharks and rays who use electrical receptors within their side lines as measuring devices. As they move in the magnetic field, the electromotive force is generated between the opposite sides of their bodies, producing an electric current in the surrounding sea water. Using electrical receptors, these fish are able to register the direction of the magnetic field of the earth, moving at a speed of only 1 cm/s. However, birds and bees who have no electrical receptors can't use this method to measure magnetic field. Besides, it would be more difficult for them to use this method because of a very large specific electric resistance of air compared to that of sea water.

Experiments of Walcott *et al.* (1988) gave the answer to the question of which of these methods is used by pigeons. The researchers suggested that homing of pigeons may be based on the magnetite crystals located in their heads. Two series of experiments to test the involvement of magnetite were performed in these experiments. The alignment of the permanent magnetic domains in birds' heads was altered by (a) demagnetizing birds, and (b) magnetizing them with a strong magnetic field. Neither of the two treatments influenced pigeons' orientation or homing in bright weather, but the results obtained under overcast skies apparently indicated that demagnetizing the birds increased the scatter of their vanishing bearings. This suggests that pigeons use a magnetic sensor for their magnetic compass in nasty weather.

To use the second method of measuring the magnetic field, permanent magnets must be present in the body of an animal. In fact, in the last 30 years tiny ferromagnetic particles have been found in many animal species, from bacteria to human beings (Kirschvink *et al.*, 1985; Kirschvink *et al.*, 1992).

Magnetites Inside Us

Ferromagnetic materials were first discovered in chitons—tiny sea mollusks that feed themselves scaling water-plants off the stones. It turned out that most of the teeth of these mollusks consist of crystalline magnetite ($FeO \cdot Fe_2O_3$), which is one of the hardest substances that form in living creatures. Incidentally, like pigeons, this mollusk demonstrates the ability to find the way back home accurately. It is assumed that it finds its way with the help of magnetic teeth.

Soon after magnetic teeth of sea mollusks had been discovered, it was found that some types of bacteria are oriented and swim mostly along the magnetic field lines even if the latter is no more than 10^{-5}T. Due to this, they were called *magnetotactic bacteria*. Ferromagnetic minerals were found inside these bacteria.

Magnetotactic bacteria several microns in size contain ferromagnetic minerals in the form of 0.1 μm-large crystals. Under certain conditions a whole colony of magnetotactic bacteria can be grown. These bacteria begin to concentrate iron inside them when the concentration of oxygen decreases in the surrounding medium. When the iron concentration in the medium becomes equal to its concentration in the backwater (1.6 mg/L), the bacteria accumulate so much iron that it can reach about 1.5% of their dry mass. Besides, each bacteria contains about 20 crystals of magnetite arranged in a chain.

Calculations showed that the magnetic moment M of such a bacteria being equal to $1.3 \cdot 10^{-15}$ J/T is enough for the orientation in the magnetic field of the Earth ($B = 5 \cdot 10^{-5}$T) at room temperature (300 K) since its energy $B \times M = 6.5 \cdot 10^{-20}$ J exceeds the corresponding heat energy $kT = 4.1 \cdot 10^{-20}$ J. Evidently, the orientation of such simple unicellular organisms in the magnetic field is a merely passive process that does not depend on what the bacteria "want." Thus, dead bacteria, as well as living ones, are oriented along the magnetic field lines of the earth.

Magnetotactic bacteria are widely spread in nature. They account for more than half of the whole bacterial plankton of most ponds, and for up to

20% of the flora of the soils. Under natural conditions, as a result of the vital activity of these bacteria, water pipes can be locked because of the accumulations of iron hydroxide produced by these bacteria. What is the role of iron ions in the life of these bacteria?

In the course of their vital activity, these bacteria accumulate hydrogen peroxide, one of the strongest oxidizers acting as poison for bacteria. In the presence of iron, the oxidizing action of the peroxide is aimed at metal rather than at bacteria. Thus, the role of iron is neutralizing hydrogen peroxide, toxic for these bacteria.

Marine, bilophotrichously flagellated, magnetotactic cocci (bacteria having two flagellar bundles on one hemisphere of the cell) swim parallel to the local magnetic field (B) in the northern hemisphere, and antiparallel to B in the southern hemisphere, which would cause them to migrate downward along the downward- and upward-inclined geomagnetic field lines in the respective hemispheres. Thus the function of magnetotaxis in magnetotactic cocci had been considered to involve facilitation of downward-directed migration for these bacteria in the water column, toward the microaerobic sediments, and away from higher oxygen concentrations at the water surface (Blackemore *et al.,* 1980). However, large numbers of cocci have recently been found to maintain their position in water columns, instead of persistently migrating down along the geomagnetic field to the bottom sediments. According to Frankel *et al.* (1997), migration in the magnetic field is determined by an aerotactic sensory system.

Magnetotactic bacteria offer a unique approach to metal accumulation and separation from water systems that are either polluted or of interest to the metal recovery industry (Bahaj *et al.,* 1994). Magnetotactic bacteria possess a magnetic dipole moment which, in conjunction with the flagellar force, makes it possible to consider the implementation of certain nonconventional separators. The orientation magnetic separator involves using low strength magnetic fields ($\sim$ 10 mT) to orientate the bacteria, and relies on their directional flagellar movement along magnetic field lines to magnetic focus. The motile, magnetotactic bacteria are accumulating at the magnetic focus from where they could be recovered.

Ferromagnetic particles containing $FeO \cdot Fe_2O_3$ originally have been found in unicellular sea-water plants Anisonema, which inhabit the salt water of the marshes at the northeastern coast of Brazil (Torres de Araujo *et al.,* 1986). The cells of these plants are rather large compared to bacteria (about 20 μm long and 12 μm wide); thus they are easy to watch through a common microscope when they are moving in a droplet of salt water. Once a magnet is brought close to this droplet, the plant cells start to migrate, some of them moving towards the north pole, others towards the south pole of the magnet. Interestingly, dead cells don't move in the magnetic field but are situated along its force lines.

To evaluate the magnetic moment of unicellular water plants, it was enough to measure the time it takes the cells to turn to 180 degrees under a rapid change of the field direction to the reverse, and to record the dependence of the turning time on the field induction value. Calculations showed that the average magnetic moment of such a plant is $6.7 \cdot 10^{-13}$ J/T, which is approximately 500 times higher than that of a typical magnetotactic bacterium. As the magnetic moment of one domain of magnetite equals $1.5 \cdot 10^{-16}$ J/T, each cell of the plant contains more than 4,000 magnetic domains that occupy about 0.2% of its volume.

The biological meaning of magnetic particles of sea-water plants isn't clear. It is well known that close to the dip equator (and the northeastern coast of Brazil is situated close to it) the force lines of the magnetic field run almost horizontally. Thus, it is assumed that migration along the lines of force enable the plants to move along the surface of the seabed far from the destructive bright light and high temperature of the surface layers of water.

Since there are so many cells, it is impossible to look for ferromagnetic materials in large animals with microscopic equipment alone. To do this, a magnetometer must be applied, a device used by geologists to determine the content of magnetic ores in rocks. The modern version of such a device consists of a wire coil immersed into liquid helium and the ammeter that measures the current in its winds. At such a low temperature the metal of the coil wire becomes a superconductor, its electric resistance falling to zero. When a permanent magnet is inserted inside the coil, electric current appears in the wire. The amperage of the current, *ceteris paribus*, will be in direct ratio to the electrical resistance of the coil. Since the coil resistance is close to zero, such devices can determine magnetic additives in the samples that contain only 10 domains of a ferromagnetic.

With the help of such a magnetometer, it was possible to measure the amount of magnetite in a single bee, which turned out to be close to 10 domains. All these magnetite crystals are located in the front part of the insect's belly. Similar measurements on the samples of different parts of a homing pigeon's body revealed large magnetic particles (up to 0.1 mm large) located in the muscles of the neck. It was noted that numerous sensitive nerve-endings are usually concentrated around these magnetic particles.

Using an ultrasensitive superconducting magnetometer, Kirschvink and his coworkers (1992) have detected the presence of ferromagnetic material in a variety of tissues from the human brain. Magnetic particle extracts from brain tissues strongly resembled those precipitated by magnetotactic bacteria and fish. These measurements implied the presence of a minimum of 5 million single-domain crystals per gram for most tissues in the brain. The presence of magnetite in human tissues has a potential implication for at least one biomedical issue that has been discussed extensively in the literature—human exposure to the 50- and 60-Hz fields produced by electric power systems and appliances (Lacy-Hulbert *et al.,* 1998). Unfortunately, without obtaining additional information about the cellular location of these particles, it is impossible to predict whether magneto-mechanical effects of this sort can be dangerous for human health.

It is generally accepted that the change of the direction of a bird's flight relative to the force lines of the magnetic field must lead to deformations of the tissues surrounding the magnetic particle. This deformation is registered by the nerve-endings and transferred to the bird's brain where the direction of the flight is analyzed.

Scientists' controversy concerning the existence of the magnetic sense may finally have been resolved by the demonstration of behavioral and neural responses to magnetic fields and the discovery of magnetic receptor cells in rainbow trout (Walker *et al.,* 1997). The researchers demonstrated behavioral responses to magnet fields by training the trout to press a bar for food when they detected a magnetic field. Walker with his colleagues have located tiny crystals of magnetite, or lodestone, in the nose. The crystals act like little compass needles and give the fish information about the earth's magnetic field, which could enable it to determine its position and the direction it needs to go when it is migrating.

Basics of Magnetic Orientation

Quite often mass media tell us the stories of mass suicides of whales and dolphins, who throw themselves ashore in different places of the ocean.

Aristotle was the first who mentioned such cases. Having failed to provide any reasoable explanation for this phenomenon, he simply admitted he didn't know why they were doing this.

Joseph L. Kirschvink and his colleagues have plotted hundreds of standings of whales and dolphins along the U.S. east coast. They found that these cetaceans tend to rush aground at spots where the earth's magnetic field is reduced by the local magnetic fields of rocks. These coastal magnetic lows are situated at the ends of long, continuous channels of magnetic minima that run for great distances along the ocean floors, suggesting that the stranded whales and dolphins were using these magnetic troughs for navigation and, having failed to see the stop sign at the beaches, ran aground. The magnetic troughs in this view are superhighways for animals with a magnetic sense. If so, the magnetic sensors of the whales and dolphins are extremely sensitive because the deepest magnetic troughs are only about 4% weaker than the background magnetic field.

Like birds, cetaceans are known to be guided in their migrations by the direction of the magnetic field lines. To find their bearings, they use magnetic particles, which have been discovered in the front part of their head. The whales are considered to swim along the magnetic field lines by memory, changing from one line to another at certain time intervals. Thus for their navigation to be successful, they need both a compass and a watch. Both their compass and their watch have been found to be magnetic. The role of the watch is played by the regular geomagnetic field fluctuations that occur in the mornings and in the evenings (remember that a similar watch is used by bees in their hive).

Besides regular geomagnetic fluctuations, irregular ones also occur. The latter are due to the solar activity, occur mostly at night, and superimposing upon a regular evening signal can interfere with and thus mask, the latter. However, if, after such a geomagnetic collapse at night, the signal in the morning is still sufficiently distinct, the whale "sets his watch to the morning" and continues moving in the right direction. Mass death of cetaceans are assumed to be happening when irregular fluctuations of the magnetic field occur in the mornings, thus hiding the regular morning signal from whales. In such cases the whales, fully relying on their magnetic navigational systems, don't realize that the morning is coming. Their calendar day becomes longer, and moving to the next magnetic field line (according to their personal map) gets delayed. As a result, when this mistake happens to occur close to the shore, the whales automatically cast themselves ashore.

The research in geomagnetic characteristics of the places of mass death of whales showed that most of the places are situated so that the seashore line is directed almost perpendicular to the magnetic field lines. Therefore, it is assumed that the reason why migrating whales cast themselves ashore is because they sometimes can't cancel the magnetic compass orientation in time to use other sensors.

The research in birds' ability to find their way led scientists to the conclusion that migrating birds can define the direction of their flight not only by their magnetic compass but also by analyzing the position of the sun or the stars in the sky (Wiltschko and Wiltschko, 1991). Evidently, birds use solar or star compasses while flying in fine weather, but when it becomes difficult to determine the position of the heavenly bodies, they resort to the magnetic mode of orientation. Which compass, then, do birds trust more? Which of the three compasses is inborn?

The idea that the birds have an inborn compass has appeared quite recently. Before that scientists used to think that birds have no reason to have an inborn compass to find the direction of the migration flight since chicks take their first flight together with adult birds, whereas later on birds can migrate by memory, using magnetic sense, solar and star compasses developed by that time.

But then how can we explain the fact that a one-year-old cuckoo, who has been reared in the foreign nest, and who thus migrates by itself, manages to determine the direction to the hibernacle correctly? It looks like every chick of a migrating bird has a genetically inborn direction of the future flight. To examine this possibility, interesting experiments were carried on pied flycatchers, migrating birds who spend the winter on the western coast of Africa (see for review Able, 1991). The hatchlings had been moved to a place where they had been kept for two and a half months until they started to fly. The direction of their first short flights was registered. The analysis showed that the average direction of these flights coincides with the direction of the place of their first prospective hibernacle (westwards).

It was supposed that nature genetically predefines the direction of the migration flight and then relegates it to only one of the navigational systems (magnetic or solar). Thus only one of the following algorithms is possible for the pied flycatcher:

1. If the inborn compass is solar, the algorithm can look like this: "At noon determine the projection of the sun to the horizon, turn the chest to this point, then turn 90° clockwise. Take this direction to move. In order not to be lost at night and in nasty weather, immediately calibrate the magnetic compass by the solar one and determine at what angle to the direction of the magnetic field lines you must fly."
2. If the inborn compass is magnetic, the program may read as follows: "Remember to fly at the angle of 90° counter-clockwise relative to the direction of the magnetic induction vector at the place of birth. In order not to be lost during magnetic storms and above geomagnetic anomalies, calibrate your solar compass by the magnetic one and find out in what direction to fly according to the sun."

To find out what algorithm nature chose and which compass is the inborn one, the hatchlings were placed for 12 days into an artificial magnetic field with the induction vector the same value as the given point of the geomagnetic field, and the direction shifted by 90 degrees to the east. Two months later the birds were released into a normal magnetic field and they started to fly at short distances in the direction of the future hibernacles. It turned out that, instead of flying to the west, they flew to the north. Clearly, they were governed by algorithm 2 (based on an inborn magnetic compass). The use of algorithm 1, evidently, should have forced the birds to fly southwards. Thus, the ability of migrating birds to find their way by the magnetic field is inborn for some of them.

Paleomagnetism and Magnetotactic Bacteria

How do we determine the age of rocks? One of the methods is measuring residual induction (the magnetic induction when the field is reduced to zero) of a rock. It is based on the fact that, while rock forms, magnetic particles in it become oriented along the magnetic field lines of the earth. In millions of years, despite the change in the direction of the earth magnetic field in a specific place, the direction of the residual magnetic induction vector of the rock remains the same. The petrified rock prevents the magnetic particles from rotating in keeping with the changes of induction vector of geomagnetic field.

The position and the speed of the movement of magnetic poles of the earth in different geological periods are well known now. For example, nowadays the speed of magnetic pole movement is about 5 km per year. Therefore, the age of a rock can be defined rather accurately by the deviation of the residual induction vector of the rock from the direction of the induction vector of the magnetic field of the earth in the place where the rock lies. Knowing the age of a rock is essential for planned search of minerals. Paleomagnetic methods (the essense of which we just described) are widely used in sea geology since the minerals lying under the seas and oceans become more accessible to us.

It was previously thought that magnetic particles contained in the sediments under the sea were brought there as a result of eolation from the continental rocks, volcanic explosions, and falling of micrometeorites. A very different view was taken by Kirschvink and Lowenstam (1979), who suggested that fossil remnants of magnetite-mineralizing organisms, in particular magnetotactic bacteria, might play an important role in the magnetization of sediments deposited in certain detritus-poor aquatic systems. The results indicated that fossil bacterial magnetite is the main carrier of natural residual magnetization in sediments ranging from Quarternary to Eocene since sometimes the age of magnetic remnants of these bacteria in the sediments can be 50 million years. Not only the remnants of prehistoric bacteria, but also their living relatives contribute to the magnetic properties of a rock. According to the estimations of the scientists, one gram of such a rock contains from 10^7 to 10^8 magnetotactic bacteria and their remnants, and their total magnetic moment can reach 10^{-6} J/T.

The revealed correlation between the magnetism of underseas and the vital activity of microorganisms opens up a new geophysics/biology interdisciplinary science. Now, animal magnetism is not something to be sniffed at. The mystery of magnetic sense and its use to guide the movement of the organisms, as small as bacteria or as large as whales, is being investigated.

10

Optima for Animals: from Mouse to Elephant

The world of animals is rich and varied. Not only is the number of different living creatures existing on earth impressive, but so is the variety of their dimensions. In fact, when we compare animals within a single class—that of terrestrial mammals—it turns out that their mass varies from 3 g for a shrew to $3 \cdot 10^6$ g for an elephant. The variety in the world of birds is not that pronounced, though still impressive: from a 1 g colibri to the 100 kg African ostriches. It is evident that dimensions and body mass are closely connected with an animal's lifestyle. What are the physical regularities that determine this relationship?

Body Mass and Lifestyle

When you watch an elephant in a zoo, it seems that all its movements are artificially slowed down. When you watch a mouse, the impression is just the opposite. The mouse moves very quickly and dexterously. It makes you think that the rhythm of life, its biological clock, depends on the mass of an animal's body. In fact, it is true. The elephant's lifetime is about 70 years but the mouse's is only two or three. The female elephant's pregnancy lasts for 18–22 months, whereas it is not more than 23 days for a mouse. It has been found that the frequency of heart contractions and breath also decreases for different mammals with the increase of the mass of the body proportional to $M^{-0.25}$ (Stahl, 1967).

The rate of the biological clock differs not only for different animal species but for the same organism while it is growing. Examples are not difficult to find. Everyone knows that babies sleep several times a day, children between 2 and 5 sleep twice a day, and adults sleep only once a day. Thus, step by step, in the course of a child's growth, the period of its biological clock goes up to 24 hours. So how can we explain the dependence of different biological processes on the mass of a mammal?

Let us try to derive the dependence that would relate the time of an animal's reaction to external irritations to its mass. It is obvious that the reaction time depends on the dimensions of an animal's body and the speed of the excitation propagation along its nerves (see Chapter 1). The speed of the nervous impulse propagation is approximately the same for different animals. Thus, the reaction time for an animal must be proportional to its linear dimensions. It is evident that for the geometrically (isometrically) similar animals, the length of the body increases with the mass M of the body proportional to $M^{0.33}$. At a constant speed of a nervous impulse propagation, the time required for the response to irritants outside and inside the animal will be proportional to its length, that is, $\sim M^{0.33}$.

Allometry of a Skeleton

Of course, real organisms can't always be considered as geometrically similar to one another. Still, some of their proportions change in a regular way as their mass grows. Biologists usually call this nonisometric change of dimensions allometric (*alloios* means "differing" in Greek). Galileo Galilei must have been the first one to notice the allometric relationships between large and small animals. In *Dialogues Concerning Two New Sciences* (1637) he points out that the bones of a huge animal, such as an elephant, have much larger dimensions than the proportionally increased bones of small animals, such as a shrew. In comparing an elephant and a shrew, whose masses differ by 10^6 times, "proportionally" implies "a hundred times."

To explain why limb bones become allometrically larger together with an animal's dimensions, we will assume that the limbs of mammals support the mass of their bodies. Then the increase of the mass of an animal has to result in the increase of the strength of the support; that is, the limb bones. To increase the strength of the support proportionally to the load, the cross-section area of the supports (the limbs) has to be proportional to the mass of the body, M. If the length of supports increases isometrically (i.e., $\sim M^{0.33}$), then the volume (and thus the mass) of the supports will be the product of their cross section and length; that is, it will be proportional to $M^{1.33}$.

In other words, comparing an animal's limbs to supports, we come to the conclusion that the mass of a skeleton, the supporting system of an animal, constitutes a larger percentage for larger animals than for smaller ones. The $M^{1.33}$ dependency would be true for the animals that consist only of limbs, and that witness only gravitational loads. In reality animals have a lot of bones that bear relatively small loads (for example, chest bones) and thus they don't need to have the cross-section area of the body directly proportional to their body mass. Besides, real animals move and thus experience not only gravitational loads, but also the forces determined by acceleration and deceleration. Evidently, for optimization, the mass of moving bones should be as low as possible. Thus, the skeleton mass of actual animals is proportional to only $M^{1.08}$ (Prange *et al.*, 1979). This is rather far from $M^{1.33}$, in which only loads due to gravity are taken into account.

Thus large and small animals, theoretically, could be isometrically similar to one another, but in reality, such relationships could hold only for immobile animals, and only in the state of weightlessness. It renders the isometric similarity of animals unrealistic. The second type of similarity is gravitational (or static), which considers only gravitational forces acting on animals. This type is more realistic, though it is still far from reality.

The third type of similarity, *elastic* similarity, was proposed by McMahon (1973). It takes into account not only gravitational forces acting on supports, but also elastic forces that appear due to the deformation of supports. It is obvious that a long and thin vertical column will function as an excellent support until, for whatever reasons, its axis deviates from the vertical. Once it happens, a column will start to bend due to its weight, and can break under a certain relation of the relevant parameters. It can be shown that the parameters that determine the strength of such long and thin columns are Young's modulus, as well as the columns' length and diameter. According to McMahon, to avoid such breaks, the length (L) and diameter (d) of the skeleton bones of an animal should depend on the body mass, M, as follows:

$$L \sim M^{1/4}, \text{ and } d \sim M^{3/8}$$

from which it follows that the bones mass should vary proportionally to M.

Stepping Frequency and Gaits

Large birds flutter their wings more seldom than do small ones. Why does a running mouse make many more steps per minute than an elephant? If they are moving in a dynamically similar way, their stride length is proportional to the leg length, L. Pennycuick (1975) measured the stride frequency of African cursorial mammals as they moved spontaneously in their natural habitat, and found it to be proportional to $L^{-0.57}$ in walking, $L^{-0.53}$ in trotting, and $L^{-0.49}$ in galloping. The general rule that larger animals need more time to make the same movement than do the smaller ones can be explained if we consider stepping movements in greater detail.

Let us look at how the position of the mass center changes at walk. Figure 10.1 schematically shows positions of a human being and his/her mass center in two sequential phases of walk. If we assume that at the moment the legs touch the ground they are straightened, it becomes clear that the position of the mass center is the lowest if both legs touch the ground. The highest position of the mass center refers to the moment when the leg that touches the ground is in the vertical position. The mass center is located in the lower part of the corpus just below the belly button; therefore, in the process of walking it moves along an arch of the circle with the radius that can be taken equal to the length of the leg.

A body moving at a speed v along a circle with the radius L has acceleration equal to v^2/L, which is directed towards the center of the circle. In the process of walking a human body is influenced by two forces, the first being the force of gravity and the second the reaction of the support. The resulting force is centripetal force, which we will concentrate on. It is evident

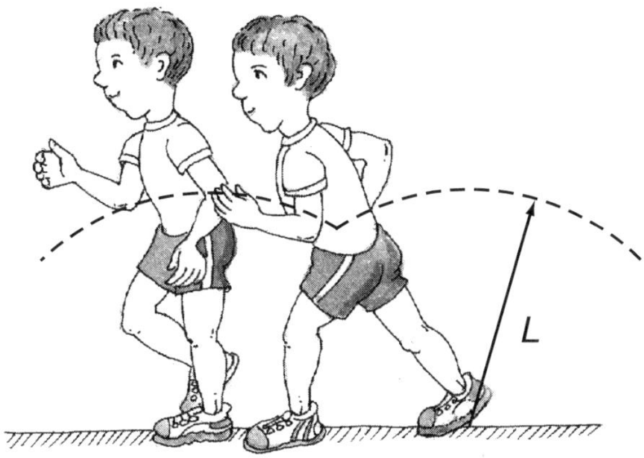

FIGURE 10.1. Schematic outlines of a man walking, showing positions of body mass (dotted line).

that its value can't exceed that of gravitational force. Therefore what is typical to walking is the inequality $mv^2/L \leq mg$, where m is the mass of a walking human being. It follows from this inequality that

$$v^2/gL \leq 1 \tag{10.1}$$

The expression in the left-hand part of the inequality (10.1) is called the Froude number. It is evident that to provide for the maximum walking speed the Froude number must be equal to 1. Let's now try to compare the movement of different mammals having equal Froude numbers. When animals travel with equal Froude numbers, their speed v is proportional to the square root of leg length, $v \sim L^{0.5}$. Therefore, the stride frequency (v/L) is proportional to $L^{-0.5}$, which is very close to the data obtained by Pennycuick (1975).

The length of an adult human's leg is about 0.9 m. Substituting this L value into (10.1) we obtain the value of 3 m/s for the maximum speed of human walking speed, which correlates with the actual value of this quantity. Children have shorter legs than adults and their maximum walking speed is lower. Therefore, children usually start running to keep up with adults.

If we are not in a hurry we stroll. Then we feel we are late and start running. When we are walking, at any moment of time at least one leg touches the ground. Unlike walking, while running there are short time intervals when we don't touch the ground. Hence this way of moving is a sequence of jumps. An adult can walk with a maximum speed of about 2.5 m/s. Running permits us to increase the speed of movement immensely. The speed of a running sportsman can be as much as 10 m/s.

Quadrupedal mammals, such as horses, walk at low speed, trot at intermediate speed, and gallop at high speed. The measurement of oxygen consumption (per distance unit) of ponies trained to walk, trot, or gallop (Hoyt and Taylor, 1981) showed that walking is most economical at a speed below 1.8 m/s, trotting is most economical between 1.8 and 4.7 m/s, and galloping is most economical at higher speeds. Humans also choose walking or running so as to minimize oxygen consumption at their current speed.

Alexander and Jayes (1983) researched the gaits of mammals ranging from small rodents to rhinoceroses and showed that there is a strong tendency for different mammals to change from symmetrical gait (a trot or pace) to asymmetrical gait (one of the forms of gallop) at Froude numbers between 2 and 3. Adult men break into running at a mean Froude number of 0.37. Other bipeds change from walking to running or hopping at about the same Froude number as human beings.

Some of these principles can be applied to legged robots (Alexander, 1989) to save energy and avoid the problems that might arise from excessive generation of heat. Comparison with animals showed that most of the robots travel with Froude numbers < 0.5; that is, at a speed at which animals of equal hip length would walk.

Jumping Performance and Body Mass

A grasshopper who is only slightly longer than a centimeter jumps as high as a locust who is five times as large. A kangaroo rat whose dimensions are the same as those of a rabbit can jump as high as an adult kangaroo. Why do the animals with a similar form jump to the same height no matter what their mass and dimensions are?

Let's take, for example, two geometrically similar animals with all the dimensions increased proportionally. We can make quite a reasonable assumption that the mass of the muscles that take part in the jumps, m_j, is the same fraction of the body mass, M, for both animals. Otherwise, their ratio $c_j = m_j/M$ wouldn't depend on the mass of the animal (Schmidt-Nielsen, 1984).

The force developed by the muscle is proportional to its cross-section area because the maximum force depends on the amount of muscular cells that occupy this area. The ratio of the maximum force of a muscle to its cross-section area varies within a very narrow range, as a rule being equal to 40–60 N/cm^2. The limiting value of the whole muscle contraction (shortening) is proportional to its length. Therefore, the work, W, performed by the muscle and equal to the product of the force and shortening must be proportional to its volume (or its mass M). It means that $W = aM$, where a is a constant whose value is equal to maximum work that can be performed by the muscle having a unit volume.

Suppose the muscle contraction at a jump assigns a speed v to a body having the mass M. To assign the speed v to the mass M it is necessary to spend the energy $Mv^2/2$. Equating this energy to the work performed by the muscle, $a \cdot c_j \cdot M$, we obtain

$$v^2 = 2a \cdot c_j \tag{10.2}$$

The maximum height reached by a vertically thrown body is known to be equal to $v^2/2g$. Then, it follows from (10.2) that geometrically similar animals can accelerate themselves to the same speed, and thus jump to the same height.

Shark and Mackerel

Everybody knows that a shark always catches a mackerel if, of course, the mackerel has no place to hide. Moreover, it catches not only a mackerel, but any small fish. But why does a large fish have a greater speed than a small one?

The resistance force that a fish has to overcome while moving in water is proportional, for the same body shape, to the cross-section area of the fish, S, and the square of its speed, v^2. Thus the expression for the power, N, that a fish uses while moving is as follows:

$$N = k_1 S v^2 v \tag{10.3}$$

where k_1 is a constant.

On the other hand, it was shown earlier that the maximum power of each muscle of an animal has to be proportional to its volume. Obviously, this must be true for the whole organism, which results in the equation:

$$N = k_2 Q \tag{10.4}$$

where Q is the volume of a fish and k_2 is a constant. Equating (10.4) to (10.3) we obtain:

$$v = k_3 (Q/S)^{1/3} \tag{10.5}$$

where k_3 is a coefficient independent of the dimensions of the fish. Let us assume that v_1, v_2, Q_1, Q_2, S_1, and S_2 are the speed, volume, and cross section of the shark and the mackerel. Then we obtain from (10.5):

$$(v_1/v_2)^3 = Q_1 S_2 / Q_2 S_1 \tag{10.6}$$

Assuming that the form of the shark and the mackerel are similar and their length is equal to L_1 and L_2 we can state that $Q_1/Q_2 = (L_1/L_2)^3$ and $S_1/S_2 = (L_1/L_2)^2$. Thus, the expression (10.6) can look different:

$$(v_1/v_2)^3 = L_1/L_2$$

Taking into consideration that a shark's body can be 5–10 m long, whereas that of a mackerel can be 0.5 m, we see that a shark will always catch a mackerel.

Carrying Loads on the Head

When travelling in Africa and Asia, many people are impressed by women bearing huge loads on their heads. Sometimes the weight of what they are carrying on top of their heads can be as much as 70% of human weight. Of course a woman can't put such a load onto her head herself, so you can quite often see two people having problems putting the load on a woman's head, and then her carrying it away easily. Is it really easier to carry a load on top of your head?

We know that the rate of a person's oxygen consumption can be considered an indicator of energy loss. One liter of consumed oxygen corresponds to the energy equal to 20.1 *kJ*. Experiments with volunteers (soldiers) showed that, with the usual way of bearing loads (on the back), oxygen consumption grows proportionally to the increase of the load's weight. A load equal to 50% of the weight of the bearer's body increases the energy loss by 50%. When the volunteers were asked to carry a load on top of the head, the dependence of oxygen consumption was the same. It was extremely surprising (Maloiy *et al.*, 1986) that African women can bear 50% loads on their heads, and the oxygen consumption will increase by 30% only. How does it happen?

Only with the help of research was the answer given. Let's look at a typical picture. A woman returns home with a full bucket on her head. However strange it might seem, not a drop is spilled on the ground. It means that in the process of walking there is no vertical component of the acceleration. Hence, the mass center of an African bearer together with the load doesn't shift in the vertical direction.

When a human being walks, the position of the mass center is influenced by pronounced vibrations. It periodically goes up and down for a few centimeters. Thus, with each step we have to spend a certain amount of energy, which is absolutely useless. This manner of movement (usual to us) can be compared with the way a new driver drives a car, trying to keep the average speed constant by pressing the accelerator or the brakes alternately, and as a result, spending more fuel.

Unlike Europeans, people in Asia and Africa worked out a manner of walking that enables them to keep the mass center in a constant position, which significantly decreases the energy loss.

Breathing-Tuned Oscillator

Baudinette and his colleagues (1987) showed that wallabies take one breath per hop throughout their range of hopping speeds. They suggested that the respiratory system is a mechanical oscillator and breathing is largely passive, driven by hopping movements. Similar suggestions have been made for horses, dogs, and jackrabbits, which take one breath per stride when they gallop.

If two pendulums hang on the same bar, the vibrations of the first pendulum after several cycles will cause the synchronous vibrations of the other one. A similar effect can be observed in the organism of a running animal. Here two pendulums interact—the periodic movements of the animal as a whole and its lungs. Figure 10.2 shows the breath pendulum of a kangaroo. On breathing in the lungs are filled with air and the mass center of the abdomen moves to the left. On breathing out it moves to the right. Elastic properties of the diaphragm and other tissues are denoted by the spring. Organs damping breath vibrations are shown as a dashpot.

The minimum energy loss at running evidently corresponds to the occasions when inertial forces appearing in the body of an animal under its periodical acceleration and deceleration facilitate (not harden) the breathing movements. Thus we can draw a conclusion that the breath frequency of an animal must be very close to the frequency of its jumps.

Specials experiments on kangaroos, horses, hares, and dogs were performed to prove this idea (for review see Alexander, 1987). It turned out that the most probable ratio of jump frequency to breathe frequency (especially for the high-speed run gallop) is 1:1. As far as humans are concerned, the correlation between the running and breathing frequencies

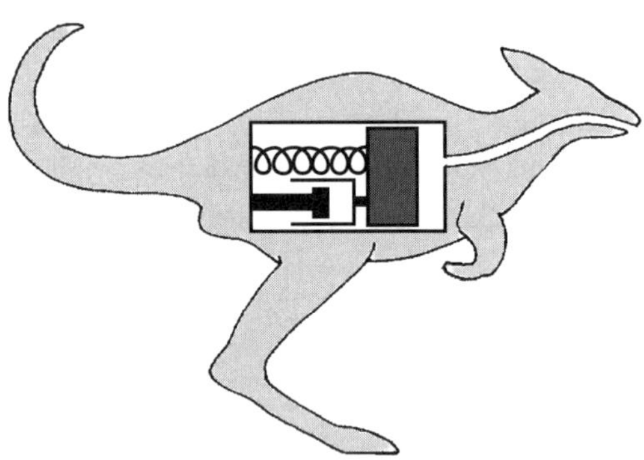

FIGURE 10.2. The hypothetical oscillator of wallabies. (Modified from Alexander, 1987. Reproduced by copyright permission of *Nature*.)

seems to be more complicated. Unlike quadrupeds, this ratio can acquire different values (4:1, 3:1, 2:1, 1:1, 5:2, 3:2) though 2:1 seems most probable.

Why running and breathing frequencies of a human are apparently independent must be because of the vertical position of the body. Hence, the movement of a human being is accompanied by vertical shifts of abdominal organs, and the horizontal forces of inertia at running influence a breathing pendulum of a human being to a much lesser extent than that of a quadruped.

Let's return to the experiment with two pendulums; we see that the forced vibrations of the second pendulum will reach the maximum value of amplitude when the periods of their self-vibrations coincide. Then it can be expected that the frequency of self-vibrations of the breathing pendulum is close to the frequency of running. It turns out to be true. Experiments performed on breathless dogs showed that the frequency of self-vibrations of breathing pendulums is about 4 Hz, and the frequency of jumps for galloping is 3.2 Hz.

The frequency of self-vibrations of a breathing pendulum has a constant value for a given animal. Hence, when the speed of running changes, the frequency of jumps must be kept constant, too. It is demanded by the minimization of energy loss. In fact, both quadruped and biped (kangaroo) runners keep the gallop frequency constant, though the length of the jumps can vary up to several times.

Energy and Body Mass

Everyone who has a hamster at home or watches it elsewhere is deeply impressed by its engorgement. In a day a hamster or any other small rodent can eat as much as it weighs. Meanwhile the mass of the everyday ration of an elephant is only 1/10 of its mass. The calorific value of an elephant's meals is about the same as the hamster's, because both of them eat only plants. What does the amount of food necessary for an animal to keep its normal vital activity depend on?

Energy supplies that we get with food is the chemical bond energy of the molecules. We will call the process involved in the release of this energy *metabolism*, and the rate at which this energy is released, the *metabolic rate*. Metabolism can be both aerobic, which means that it can occur only under a continuous oxygen income into the organism, and anaerobic, which doesn't demand any oxygen income. Aerobic metabolism provides much more energy for the organism from the same amount of food compared to the anaerobic one, thus covering the most of our energy loss. An organism needs more oxygen when it is turning from a state of rest to locomotive activity. Warm-blooded animals consume more oxygen than cold-blooded ones because in the relatively cold medium they constantly lose heat because of convection and emission. Therefore, if we measure the consumption of oxygen by an animal we can evaluate with sufficient accuracy its metabolic rate because it has been calculated that the consumption of 1 cc of oxygen is accompanied by the production of 20 J of heat energy. The amount of produced heat doesn't depend on the type of food.

In his very famous paper *Body size and metabolism*, written more than 60 years ago, M. Kleiber showed that the basal, or resting, metabolic rate for animals ranging from mice to elephants scales closely as $M^{0.75}$. Kleiber came to the conclusion that the data obtained can be best described by the equation

$$P_{\text{met}} = 70M^{0.75} \tag{10.7}$$

where P_{met} is the metabolic rate (in kcal/day) and M is the body mass of mammals in kg.

To solve the problem of relative engorgement of a hamster compared to an elephant we simply need to calculate the metabolic rate for a unit weight of a mammal by dividing both parts of equation (10.7) by M:

$$p_{\text{met}} = 70M^{-0.25} \tag{10.8}$$

where p_{met} is the specific metabolic rate (in kcal·day^{-1}·kg^{-1}). Now let's compare p_{met} for a hamster and an elephant, whose masses differ by 100,000 times. It becomes evident that energy demands of one gram of a hamster exceed those of an elephant by more than 17 times ($100\,000^{-0.25} = 17.8$).

Let us try to explain the dependence (10.8) between the specific metabolic rate and the mass of an animal; the starting point will be the fact that constant body temperature of warm-blooded animals is supported by heat production under metabolic processes. Let us suppose an animal has the shape of a sphere with the radius R, and the unit mass of the animal every second releases p_{met} of heat energy. Then the amount of heat, Q, that is released in the organism every second as a result of metabolic processes becomes equal to

$$Q = (4/3)\pi R^3 p_{met} \tag{10.9}$$

Since body temperature remains constant, the amount of heat energy emerging in the organism as a result of metabolic processes must equal the amount of heat transferred from an animal to the medium. The amount of heat, Q, which is transferred from a warmer body to a colder one per time unit, is proportional to the contact area, S, difference of the bodies' temperatures, ΔT, heat conductance of the medium between the bodies, ψ, and is inversely proportional to the thickness of the layer of the medium, Δx, thus

$$Q = S\psi(\Delta T/\Delta x)$$

If any changes of our spherical animal's dimensions don't affect the quantities ψ and $\Delta T/\Delta x$, then

$$Q = 4\pi R^2 k, \tag{10.10}$$

where $k = \psi(\Delta T/\Delta x)$. Equating the amount of heat released in the organism under metabolic processes (10.9) with the heat emitted by the body through its surface (10.10), we obtain

$$p_{met} = 3kR^{-1} \tag{10.11}$$

As $M = (4/3)\pi R^3$ then expressing R in (10.11) by means of M we finally get the following expression:

$$p_{met} = 3k(4\pi/3)^{1/3} M^{-1/3} \sim M^{-1/3} \tag{10.12}$$

Therefore, the analysis of heat energy balance for similar animals leads to the conclusion that the amount of oxygen consumed by a mass unit of an animal must decrease in inverse ratio to the value of this mass. Theoretical (10.12) and experimental (10.8) estimations of p_{met} differ because as a rule larger animals are not similar to smaller ones. According to the principle of elastic similarity, their dimensions vary in direct ratio to $M^{1/4}$, not to $M^{1/3}$ (McMahon, 1973).

Specific metabolic rate is the most important indicator of how fast the time flies for a given animal. As time is in inverse ratio to rate, it follows from (10.8) that the metabolic time constant, t_{met}, depends on the body mass according to the ratio

$$t_{\text{met}} \sim M^{1/4} \tag{10.13}$$

The term "metabolic time" can be applied to nearly all the processes within a living organism. One example is enough to illustrate this issue. Everyone knows that pills recommended by a doctor must be taken several times a day, otherwise they are ineffective. It is necessary to take a drug at certain intervals to maintain its constant concentration in blood due to its decomposition in the organism.

Figure 10.3 shows the change of the concentration of pharmaceuticals in human or animal blood after a single introduction. The decrease of the drug concentration can be approximated by an exponent, the order of which contains time normalized by τ_d, a constant quantity having the dimension of

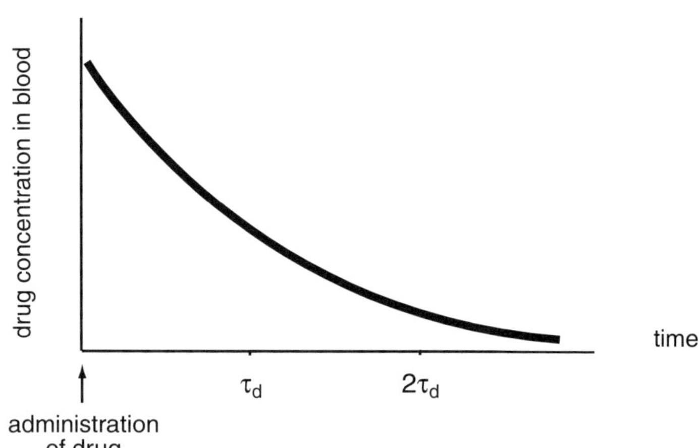

FIGURE 10.3. Decrease in blood concentration of drug after its administration.

time. The time coefficient of drug decomposition in the organism, τ_d, determines the rate of its concentration decrease after a single introduction.

The rate of the drug decomposition in the organism must be proportional to the metabolic rate. Therefore, we can assume that this biological time constant must be proportional to $M^{1/4}$. Such dependence of τ_d on $M^{1/4}$ was revealed when animals whose mass varies by several thousand times were studied (Dedrick *et al.,* 1970).

We know about a tragedy that happened because the dependence between τ_d and M was not studied enough. There is a certain drug, LSD, which is immensely interesting from the scientific point of view, to psychiatrists and neurophysiologists, which makes normal people hallucinate in a most peculiar way. A group of researchers decided to study how an elephant will react to this drug. For that, they took the amount of LSD they knew could enrage a cat and multiplied it by the ratio of the elephant and cat weights. They thought that the dose of the drug should be in direct ratio to the mass of an animal. The introduction of such a dose of LSD killed the elephant in five minutes. The researchers decided that elephants are more sensitive to this particular drug. Later, in a review this work was called an "elephantine mistake" made by the authors of the experiment. As it follows from (10.8), a dose of LSD per one kilogram should have been more than five times *decreased* for the elephant weighing about 1000 times more than a cat ($1000^{0.25} = 5.6$).

Living Wheel ?

We know now that in the world that surrounds us we can see a lot of things invented by people (a flight in the air, underwater swimming, jet planes, a parachute). Why is there no wheel in nature? Why do horses run rather than roller-skate? Why is there no fish with a propeller instead of a tail?

All rotating systems have two main advantages. First, they replace sliding friction by rolling friction, which as a rule is lower. Second, in the course of moving, their kinetic energy doesn't fluctuate, but remains constant. If we evaluate how effective a transportation means is by the ratio of its mass to the energy necessary to overcome a distance unit, than the effectiveness of a bicycle rider's movement is 1.6 kgm/J. Even someone whose legs are paralyzed and who moves in a wheel chair moves more effectively

(0.42 kgm/J) than another one who walks. Why don't the animals use a wheel then?

Because a living wheel that needs constant blood input from outside is impossible, although some exceptions are possible. For example, rotating flagella of the bacteria are protein fibers (0.2 microns in diameter) that don't need to be supplied with oxygen or nutrients. Bacteria swim screwing their helical flagella through the external medium at a speed up to 25 μm/s.

The bacterial flagellar motor is a rotary engine (for review see Elston and Oster, 1997) that derives its energy from an electrochemical gradient established between the cell cytoplasm and the periplasmic lumen. In most bacteria this gradient is set up by proton pumps. Kinetic studies suggest that at normal swimming conditions of $\sim$ 100 Hz, $\sim$ 1200 protons pass through the motor per revolution, or $\sim$ 100 000 protons/s. Although a complete molecular structure of the flagellar motor is not yet available, it consists of the rotor with a radius of $\sim$ 20–25 nm and 8–16 torque-generating stator elements. It was calculated that the torque developed by each stator element is $\sim$ 30–60 pN·nm.

So why don't animals move on wheels or screw-propellers made of biological materials that don't demand constant income of nutrients like bones? Can a wolf travel through the woods on roller-skates? Of course not. All the advantages of moving with the help of wheels disappear once we turn from a highway to off-road. It happens because, first, the sliding friction begins to increase (e.g., it is 10 times as high for sand compared to concrete), and second, the outcrops limit the minimum size of a wheel, because the maximum height of an obstacle shouldn't exceed half a radius of the wheel. Hence it becomes clear why animals have no wheels: they simply don't need them.

References

Able, K. P. 1991. "The development of migratory orientation mechanisms," *EXS* **60**:166–79.

Alexander, R. McN. 1987. "Wallabies vibrate to breath," *Nature* **328**: 477.

Alexander, R. McN. 1989. "Optimization and gaits in the locomotion of vertebrates," *Physiological Reviews* **69**: 1199–1227.

Alexander, R. McN. and H. C. Bennet-Clark. 1977. "Storage of elastic strain energy in muscle and other tissues," *Nature* **265**: 114–117.

Alexander, R. McN. and A. S. Jayes. 1983. "A dynamic similarity hypothesis for the gaits of quadrupedal mammals," *J. Zool. Lond.* **201**: 135–152.

Arnold, W. W. and U. Zimmermann. 1988. "Electro-rotation: development of a technique for dielectric measurements on individual cells and particles," *J. Electrostat.* **21**: 151–191.

Arny, D. C., S. E. Lindow, and C. D. Upper. 1976. "Frost sensitivity of *Zea mays* increased by application of *Pseudomonas syringae*," *Nature* **262**: 282–284.

Ashmore, F. Mammano, J. E. Gale, and M. Tunstall. 1995. *Active Hearing* (Flock, A., Ottosen, D., and Ulfendahl, M., Eds.), Pergamon, 337–348.

August, P.V. *et al.* 1989. "Magnetic Orientation in a Small Mammal, Peromyscus Leucopus," *Journal of Mammalogy* **70**:1.

Bahaj, A. S., I. W. Crowdace, and P. A. B. James. 1994. *IEEE Trans. Magn.* **30**: 4707–4709.

Baudinette, R. V., B. J. Gannon, W. B. Runciman, S. Wells, and J. B. Love. 1987. *J. Exp. Biol.* **129**: 251–263.

Baylor, D. A., T. D. Lamb, and K. W. Yau. 1979. "Responses of retinal rods to single photons," *J Physiol (Lond)* **288**:613–634.

Becker, F. F., X. B. Wang, Y. Huang, R. Pethig, J. Vykoukal, and P. R. C. Gascoyne. 1995. "Separation of human breast cancer cells from blood by differential dielectric affinity," *Proc. Natl. Acad. Sci. USA* **92**: 860–864.

Blakemore, R. P., R. B. Frankel, and A. J. Kalmijn. 1980. "South-seeking bacteria in the southern hemisphere," *Nature* **286**: 384–385.

Blauert, J. 1997. *Spatial Hearing*. MIT Press, Cambridge, MA.

Bownds, M. D. 1981. "Molecular mechanisms of visual transduction," *Trends In NeuroSciences* **4**: 214–217.

Braekevelt, C. R. 1994a. "Fine structure of the tapetum lucidum in the short-tailed stingray (Dasyatis brevicaudata)," *Histol. Histopathol.* **9**(3):495–500.

Braekevelt, C. R. 1994b. "Fine structure of the choroidal tapetum lucidum in the Port Jackson shark (Heterodontus phillipi)," *Anat. Embryol. (Berl)* **190**(6):591–596.

Brownell, W. E. and A. S. Popel, 1998. "Electrical and mechanical anatomy of the outer hair cell," *Psychophysical and Physiological Advances in Hearing*, edited by A.R. Palmer, A. Rees, A.Q. Summerfield, R. Meddis. London: Whurr Publishers.

Burkard, M. E. and H. D. van Liew. 1995. "Effects of physical properties of the breathing gas on decompression-sickness bubbles," *J. Appl. Physiol.* **79**(5): 1828–1836.

Butler, P. J. and D. R. Jones. 1997. "Physiology of diving of birds and mammals," *Physiol. Reviews* **77**: 837–899.

Clark, L. C. Jr. and F. Gollan. 1966. "Survival of mammals breathing organic liquids equilibrated with oxygen at atmospheric pressure," *Science* **152**:1755–1756.

Clements, J. A. 1997. "Lung surfactant: a personal perspective," *Ann. Rev. Physiol.* **59**: 1–21.

Collett, T. S. 1987. "Insect maps," *TiNS* **10**(4):139–141.

Dallos, P., A. N. Popper, and R. R. Fay, 1996. "The cochlea," *Springer Handbook of Auditory Research*, Vol. 8. Springer Verlag, New York.

Davies, A., G. F. Schertler, B. E. Gowen, and H.R. Saibil. 1996. "Projection structure of an invertebrate rhodopsin," *J. Struct. Biol.* **117**(1): 36–44.

Dedrick, R. L., K. B. Bischoff, and D. S. Zakharko. 1970. "Interspecies correlations of plasma concentration history of methotrexate," *Cancer Chemother. Rep.* **54**: 95–101.

DeVries, A. L. 1988. "The role of glycopeptides and peptides in the freezing avoidance of Antarctic fishes," *Comp. Biochem. Physiol. B Comp. Biochem.* **90B**: 611–621.

Duck, F. A. *Physical Properties of Tissue: A Comprehensive Reference Book.* Academic Press, London. 1990.

Dwyer, T. M., D. J. Adams, and B. Hille. 1980. "The permeability of the endplate channel to organic cations in frog muscle," *J. Gen. Physiol.* **75**: 469–492.

Elliott, E. A. and S. V. Dawson. 1977. "Test of wave-speed theory of flow limitation in elastic tubes," *J. Appl. Physiol.* **43**(3):516–522.

Elston, T. C. and G. Oster, 1997. "Protein turbines1: the bacterial flagellar motor," *Biophys. J.* **73**: 703–721.

Engelhardt, H. and E. Sackmann. 1988. "On the measurement of shear elastic moduli and viscosities of erythrocyte plasma membranes by transient deformation in high frequency electric fields," *Biophys. J.* **54**(3): 495–508.

Fazzalari, N. L. and I. H. Parkinson. 1998. "Fractal properties of cancellous bone of the iliac crest in vertebral crush fracture," *Bone* **23**:53–57.

Feeney, R. E. and Y. Yeh. 1993. "Antifreeze proteins: properties, mechanism of action, and possible applications," *Food Technol.* Jan.: 82–89.

Flamarique, N. I. and C.W. Hawryshyn. 1997. "Is the use of underwater polarized light by fish restricted to crepuscular time periods?" *Vision Res.* **37**(8):975–989.

Frank, R. S. and R. M. Hochmuth. 1988. "The influence of red cell mechanical properties on flow through single capillary-sized pores," *J. Biomech. Eng.* **110**: 155–156.

Frankel, R. B., D. A. Bazylinski, M. S. Johnson, and B. L. Taylor. 1997. "Magneto-aerotaxis in marine coccoid bacteria," *Biophys. J.* **73**: 994–1000.

Freeman, S. A., M. A. Wang, and J. C. Weaver. 1994. "Theory of electroporation of planar bilayer membranes: predictions of the aques area, change in capacitance, and pore-pore separation," *Biophys. J.* **67**: 42–56.

Fuhrman, F. A. 1986. "Tetrodotoxin, Tarichatoxin, and Chiriquitoxin: historical perspectives," *Ann. N.Y. Acad. Sci.* **479**: 1–14.

Funnell, W. R. J. 1996. "Low-frequency coupling between eardrum and manubrium in a finite-element model," *J. Acoust. Soc. Am.* **99**(5): 3036–3043.

Gabelnick, H. L. and M. Litt, Eds. 1973. *Rheology of biological systems.* Springfield: Ch. C. Thomas Publisher.

Gibson, L. and M. Ashby. *Cellular Solids: Structure and Properties.* Pergammon Press, Oxford. 1988.

Goldman, D. E. 1943. "Potential, impedance, and rectification in membranes," *J. Gen. Physiol.* **27**: 37–60.

Goldstein, S. 1987. "The Mechanical Properties of Trabecular Bone: Dependence on Anatomic Location and Function," *J. Biomech.* **20**:1055–1061.

Gould, J. L. 1986. "The locate map of honey bees: Do insects have cognitive maps?" *Science* **232**: 861–863.

Gross, D. and W. S. Williams, 1982. "Streaming potential and the electromechanical response of physiologically moist bone," *J. Biomech.* **15**: 277–295.

Hall, S. J., J. A. McGuigan, and M. J. Rocks. 1991. "Red blood cell deformability in sudden sensorineural deafness: another aetiology?" *Clin. Otolaryngol.* **16**(1): 3–7.

Hardung, V. 1962. "Propagation of pulse waves in visco-elastic tubings," *Handbook of Physiology; Section 2 : Circulation* (ed. by W. F. Hamilton) 107–135. American Physiological Society, Washington, D.C.

Hargitay, B. and W. Kuhn. 1951. *Z. Elektrochem. Angew Phys. Chem.* **55**:539.

Hibbeler, R. C. 1994. *Mechanics of materials*, 2nd ed. Macmillan, NY.

Hille, B. 1992. *Ionic Channels of Excitable Membranes*, 2nd ed. Sinauer Associates Inc., Sunderland, MA.

Hodgkin, A. L. and B. Katz. 1949. "The effect of sodium ions on the electrical activity of the giant axon of the squid," *J. Physiol. (London)* **108**: 37–77.

Horvath, G. and D. Varju. 1995. "Underwater refraction-polarization patterns of skylight perceived by aquatic animals through Snell's window of the flat water surface," *Vision Res.* **35**(12):1651–1666.

Hoyt, D. F. and C. R. Taylor. 1981. "Gait and the energetics of locomotion in horses," *Nature* **292**: 239–240.

Huang, Y., X-B. Wang, J. A. Tame, and R. Pethig. 1993. "Electrokinetic behaviour of colloidal particles in travelling electric fields: Studies using yeast cells," *J.Phys.D: Appl.Phys.* **26**: 1528–1535.

Hubbard, A. 1993. "A traveling-wave amplifier model of the cochlear," *Science* **259**: 68–71.

Innovation. 1993. Volume 1, Number 5 October November. "Shuttle Tile Shows Promise as Bone Replacement."

Iwasa, K. H. 1994. "A Membrane Motor Model for the Fast Motility of the Outer Hair Cell," *J. Acoust. Soc. Am.* **96**:2216–2224.

Jaeger, M. J., U. H. Kurzweg, and M. J. Banner. 1984. "Transport of gases in high-frequency ventilation," *Crit. Care Med.* **12**(9):708–710.

Kaler, K. V. I. S. and T. B. Jones. 1990. "Dielectrophoretic spectra of single cells determined by feedback-controlled levitation," *Biophys. J.* **57**: 173–182.

Keaveny, T. M. and W. C. Hayes. 1993. *J. Biomech. Eng.* **115**:534–542.

Kelly, P. J. and Bronk, J. T. 1990. "Venous Pressure and Bone Formation," *Microvascular Research* **39**: 364–375.

Kemp, D.T. 1978. "Stimulated acoustic emissions from within the human auditory system," *J. Acoust. Soc. Am.* **64**(5):1386–1391.

Kenward, K. D., M. Altschuler, D. Hildebrand, and P. L. Davies. 1993. "Accumulation of type I fish antifreeze protein in transgenic tobacco is cold-specific," *Plant Mol. Biol.* **23**: 377–385.

Ker, R. F., M. B. Bennet, S. R. Bibby, R. C. Kester, and R. McN. Alexander. 1987. "The spring in the arch of the human foot," *Nature* **325**: 147–149.

Kirschvink, J. L., D. S. Jones, and B. J. MacFadden. 1985. *Magnetite Biomineralization and Magnetoreception in Organisms: A new Biomagnetism*. Plenum, NY.

Kirschvink, J. L., A. Kobayashi-Kirschvink, and B. J. Woodford. 1992. "Magnetite biomineralization in the human brain," *Proc. Natl. Acad. Sci. USA* **89**: 7683–7687.

Kirschvink, J. L. and H. A. Lowenstam. 1979. *Earth planet. Sci. Lett.* **44**: 193–204.

Kirschvink, J. L. *et al.* 1992. "Discrimination of low-frequency magnetic fields by honeybees: biophysics and experimental tests," *Soc. Gen. Physiol. Ser.* **47**:225–240.

Kitchener, A. 1991. *The Natural History of the Wild Cats*. Comstock Publishing Associates, Ithaca, NY.

Knight, C. A., C. C. Cheng, and A. L. DeVries. 1991. "Adsorption of alpha-helical antifreeze peptides on specific ice crystal surface planes," *Biophys. J.* **59**: 409–418.

Knight, J. 1998. "Life on ice," *New Scientist* **2132**: 24.

Kuhn, W. and K. Ryffel. 1942. *Hoppe-Seyler's Z. Physiol. Chem.* **276**: 145.

Kunze, P. 1979. "Apposition and superposition eyes." Autrum, H. (ed.), *Comparative physiology and evolution of vision in invertebrates, A: Invertebrate photoreceptors*. Handbook of Sensory Physiology VII/6A 441–502. Springer, Berlin.

LaBarbera, M. 1990. "Principles of design of fluid transport systems in zoology," *Science* **249**: 992–1000.

Lacy-Hulbert, A., J. C. Metcalfe, and R. Hesketh. 1998. "Biological responses to electromagnetic fields," *FASEB J.* **12**: 395–420.

Ladak, H. M. and Funnell, W. R. J. 1993. "Finite-element modelling of the reconstructed cat middle ear," *Proc. 19th Can. Med. & Biol. Eng. Conf.*: 370–371.

Laforge, H., M. R. Sadeghi, and M. K. Seguin. 1986. "Magnetostatic field effect: stress syndrome pattern and functional relation with intensity," *J. Psychol.* **120**(3): 299–304.

Land, M. F. 1965. "Image formation by a concave reflector in the eye of the scallop, Pecten maximus," *J. Physiol (Lond).* **179**(1):138–153.

Land, M. F. 1980. "Compound eyes: old and new optical mechanisms," *Nature* **287**: 681–686.

Layton, H. E. 1986. "Distribution of Henle's loops may enhance urine concentrating capability," *Biophys. J.* **49**: 1033–1040.

Leake, C. D. 1962. "The historical development of cardiovascular physiology," *Handbook of Physiology Section 2: Circulation* (ed. by W. F. Hamilton) 11–22. American Physiological Society, Washington, D.C.

Leicester, H. M. and H. S. Klickstein. 1952. *A Source Book in Chemistry* 1400–1900. McGraw Hill, NY.

Lespessailles, E., J. P. Roux, C. L. Benhamou, M. E. Arlot, E. Eynard, R. Harba, C. Padonou, and P. J. Meunier. 1998. "Fractal analysis of bone texture on os calcis radiographs compared with trabecular microarchitecture analyzed by histomorphometry," *Calcif. Tissue Int.* **63**:121–125.

Longo, M. L., A. M. Bisagno, J. A. N. Zasadzinski, R. Bruni, and A. J. Waring. 1993. "A function of lung surfactant protein SP-B," *Science* **261**: 453–456.

Lythgoe, J. N. 1979. *The ecology of vision*. Claredon Press, Oxford, England.

Majumdar, S. and H. K. Genant. 1995. "Magnetic resonance imaging in osteoporosis," *Eur. J. Radiol.*, **20**, 193–197.

Maloiy, G. M. O., N. C. Heglund, L. M. Prager, G. A. Cavagna and C. R. Taylor. 1986. "Energetic cost of carryng loads: have African women discovered an economic way?" *Nature* **319**: 668–669.

McMahon, T. 1973. "Size and shape in biology," *Science* **179**: 1201–1204.

Muntwyler, E. 1968. *Water and Electrolyte Metabolism and Acid-Base Balance*. St. Louis, Mosby.

Murray, C. D. 1926. "The physiological principle of minimum work. I. The vascular system and the cost of blood volume," *Proc. Natl. Acad. Sci. USA* **12**: 207–214.

Nash, G. B. 1990. "Filterability of blood cells: methods and clinical applications," *Clinical Hemorheology* **10**: 353–362.

Newman, E. A. and P. H. Hartline. 1981. "Integration of visual and infrared information in bimodal neurons in the rattlesnake optic tectum," *Science* **213** (4509):789–791.

Nilsson, D. E. 1983. "Evolutionary links between apposition and superposition optics in crustacean eyes," *Nature* **302**: 818–821.

Nilsson, D. E. 1988. "A new type of imaging optics in compound eyes," *Nature* **332**: 76–78.

Nobili, R. and F. Mammano. 1996. *J. Acoust. Soc. Am.* **99**: 2244–2255.

Notter, R. H., H. Holcomb, and R. D. Mavis. 1980. "Dynamic surface properties of phosphatidylglycerol-dipalmitoyl phosphatidylcholine mixed films," *Chem. Phys. Lipids* **27**:305–319.

Pennycuick, C. J. 1975. "On the running of the gnu and other animals," *J. Exp. Biol.* **63**: 775–799.

Permutt, S., W. Mitzner, and G. Weinmann. 1985. "Model of gas transport during high-frequency ventilation," *J. Appl. Physiol.* **58**(6):1956–1970.

Pethig, R. 1991. "Dielectric properties of tissue," In: *Encyclopedia of Human Biology* (ed. by R. Dulbecco, Academic Press Inc), pp. 9–15.

Pitts, R. F. 1974. *Physiology of the Kidney and Body Fluids*, (3rd ed.) Year Book, Chicago.

Pohl, H. A. 1978. *Dielectrophoresis*. Cambridge University Press, Cambridge, UK.

Popper, A. N. and Fay, R. R. Eds. 1995. *Hearing by Bats*. Springer Verlag, NY.

Poulain, F. R. and Clements, J. A. 1995. "Pulmonary surfactant therapy," *West J. Med.* **162**: 43–50.

Prange, H. D., J. F. Anderson, and H. Rahn. 1979. "Scaling of skeletal mass to body mass in birds and mammals," *Amer. Nat.* **113**: 103–122.

Radford, E. J. 1954. "Method for estimating respiratory surface area of mammalian lungs from their physical characteristics," *Proc. Soc. Exp. Biol. Med.* **87**: 58–61.

Reilly, D. T. and A. H. Burstein. 1975. "The elastic and ultimate properties of compact bone tissue," *J. Biomech* **8**(6):393–405.

Sadowski, R. 1996. "Liquid Ventilation: back to the future," *J. Resp. Care Practitioners* **9**:35–41.

Sauer, F. A. 1985. "Interaction forces between microscopic particles in an external electromagnetic field," *Interaction between Electromagnetic Field and Cells.* A. Chiabrera, C. Nocolini, and H. P. Schwan, eds. 181–202 Plenum Publishing Corp., New York.

Schmidt-Nielsen, B. and R. O'Dell. 1961. "Structure and concentrating mechanism in the mammalian kidney," *Am. J. Physiol.* **200**: 1119–1124.

Schmidt-Nielsen, K. 1984. *Scaling. Why is animal size so important?* Cambridge University Press, Cambridge.

Schmidt-Nielsen, K. 1994. "About curiosity and being inquisitive," *Ann. Rev. Physiol.* **56**: 1–12.

Secomb, T. W., and R. Hsu. 1996. "Analysis of red blood cell motion through cylindrical micropores: effects of cell properties," *Biophys. J.* **71**: 1095–1101.

Spadaro, J. A. 1997. "Mechanical and electrical interactions in bone remodeling, *Bioelectromagnetics*" **18**(3):193–202.

Stahl, W. R. 1967. "Scaling of respiratory variables in mammals," *J. Appl. Physiol.* **22**: 453–460.

Stephenson, J. L. 1973. "Concentrating engines and the kidney," *Biophys. J.* **13**: 512–545.

Storey, K. B. and J. M. Storey. 1992. "Natural freeze tolerance in ectothermic vertebrates," *Ann. Rev. Physiol.* **54**: 619–637.

Taber, L. A. 1998. "An optimization principle for vascular radius including the effects of smooth muscle tone," *Biophys. J.* **74**: 109–114.

Talmadge, C. L., A. Tubis, G. R. Long, and P. Piskorski. 1998. "Modeling otoacoustic emission and hearing threshold fine structures," *J. Acoust. Soc. Am.* **104**(3 Pt 1):1517–1543.

Taneva, S. and K. M. W. Keough, 1994. "Dynamic surface properties of pulmonary surfactant proteins SP-B and SP-C and their mixtures with dipalmitoylphosphatidylcholine," *Biochemistry* **33**: 14660–14670.

Torres de Araujo, F. F., M. A. Pires, R. B. Frankel, and C. E. M. Bicudo. 1986. "Magnetite and magnetotaxis in algae," *Biophys. J.* **50**: 375–378.

Ulfendahl, M. 1997. "Mechanical responses of the mammalian cochlea," *Prog. Neurobiol.* **53**: 331–380.

van Liew, H. D. and S. Raychaudhuri. 1997. "Stabilized bubbles in the body: pressure-radius relationships and the limits to stabilization," *J. Appl. Physiol.* **82**(6): 2045–2053.

Vogt, K. 1975. *Z. Naturforsch.* **30c**: 691.

von der Emde, G., S. Schwarz, L. Gomez, R. Budelli, and K. Grant. 1998. "Electric fish measure distance in the dark," *Nature* **395**: 890–894.

von Frisch, K. 1949. "The polarisation of skylight as a means of orientation during the bee's dances," *Experientia (Basel)* 5, 142–148.

von Frisch, K. 1967. *The Dance Language and Orientation of Bees.* Harvard University Press.

von Nahmen, A., M. Schenk, M. Sieber, and M. Amrein. 1997. "The structure of a model pulmonary surfactant as revealed by scanning force microscopy," *Biophys. J.* **72**: 463–469.

von Neergaard, K. 1929. "Neue Auffasungen uber einen Grunbegriff der Atemmechanik," *Z Gesamte. Exp. Med.* **66**: 373–394.

Walcott, C., J. L. Gould, and A. J. Lednor. 1988. "Homing of magnetized and demagnetized pigeons," *J. Exp. Biol.* **134**:27–41.

Walker, J. D. 1975. "Karate strikes," *Am. J. Phys.* **43**(10): 845–849.

Walker, M. M. *et al.* 1997. "Structure and function of the vertebrate magnetic sense," *Nature* **390**: 371–376.

Walker, M. M. 1997. "Magnetic orientation and the magnetic sense in arthropods," *EXS* **84**:187–213 .

Walsh, C., P. A. Sullivan, J. S. Hansen, and L.W. Chen. 1995. "Measurement of wall deformation and flow limitation in a mechanical trachea," *J. Biomech. Eng.* **117**(1):146–152

Weinstein, A. M. 1994. "Mathematical models of tubular transport," *Ann. Rev. Physiol.* **56**: 13–45.

Wiltschko, W, and R. Wiltschko. 1991. "Orientation in birds. Magnetic orientation and celestial cues in migratory orientation," *EXS* **60**:16–37.

Wilson, E. O. 1972. "Animal Communication," *Scientific American*, **227**(3): 53–60.

Wiltschko, W. and R. Wiltschko. 1996. "Magnetic orientation in birds," *J. Exp. Biol.* 199 (Pt. 1): 29–38.

Yack, J. E. and J. H. Fullard. 1993. "What is an insect ear?" *Ann. Entomol. Soc. Am.* **86**:677–682.

Yager, D. D. and R. R. Hoy. 1986. "The cyclopean ear: a new sense for the praying mantis," *Science* **231**: 727–729.

Yanagihara, N., H. Von Leden, and E. Werner-Kukuk. 1966. "The physical parameters of cough: the larynx in a normal single cough," *Acta Otolaryngol (Stockh)*. **61**(6):495–510.

Yonemori, K., S. Matsunaga, Y. Ishidou, S. Maeda, and H. Yoshida. 1996. "Early effects of electrical stimulation on osteogenesis," *Bone* **19**: 173–180.

Zhuang, H., W. Wang, R. M. Seldes, A. D. Tahernia, H. Fan, and C.T. Brighton. 1997. "Electrical stimulation induces the level of TGF-beta1 mRNA in osteoblastic cells by a mechanism involving calcium/calmodulin pathway," *Bioch. Bioph. Res. Comm.* **237**: 225–229.

Zierler, K. L. 1958. "A simplified explanation of the theory of indicator-dilution for measurement of fluid flow and volume and other distributive phenomena," *Bull. John Hopkins Hosp.* **103**: 199.

INDEX